PRAISE FOR *Chasing Space*

"For someone of such remarkable a̶ strates impressive humility and d̶ cusses overcoming myriad setback about his mistakes as well as his pursue their dreams with determinatiᴏ̶ᴜ ᴀ̶

—*Publishers Weekly* (starred review)

"Though Melvin refers to his story as one of grit and grace, he does so without a shred of ego. . . . Second chances played a key role in Melvin's ability to overcome obstacles and achieve his many dreams, and he'll have readers wondering how many chances it woud take them to do the same." —*Booklist*

"As a black woman researcher at NASA in the 1950s, I found *Chasing Space* very revealing. While I performed the calculations that assisted men on their missions to the moon, I always hoped that in my lifetime I would see a black man fly into space. Not only did this book make me feel as if I were on this journey with Leland, through his eyes, it reminded me of how small the planet really is, and therefore, how insignificant our differences as humans are—the kind of issues that kept my own legacy at NASA hidden for so long. *Chasing Space* is a must-read—a flight you shouldn't miss."

—Katherine Johnson, former NASA physicist and mathematician, and recipient of the Presidential Medal of Freedom

"Nerd-kid almost burns down his house with a chemistry experiment gone bad, then grows up and plays for the National Football League, then flies to the International Space Station. That actually happened, and it is quite a story. In *Chasing Space*, Leland Melvin, the former NASA astronaut, shares this remarkable trajectory of his life. En route he tackles stupendous obstacles with dogged determination, showing you what is indeed possible in life—if you believe."

—Neil deGrasse Tyson, astrophysicist at the American Museum of Natural History and author of *Astrophysics for People in a Hurry*

"Henry Ford said, 'If you think you can or think you can't, you're right.' Leland Melvin proves it. His story inspires me and reminds me of the value of staying positive and pushing through adversity. *Chasing Space* is uplifting and motivating."

—Simon Sinek, optimist and author of
Start with Why and *Leaders Eat Last*

"A story of passion and perseverance told by a humble hero. Insightful and inspiring!"

—Angela Duckworth, *New York Times* bestselling author of *Grit*

"Leland's story moves fast from the deck of his homebuilt skateboard to the flight deck of the space shuttle. He shows us how he made the best of every opportunity and every setback on his way from his hometown to outer space. It moves fast; once you get started you'll want to join the chase."

—Bill Nye, chief executive officer of The Planetary Society

"Leland has helped me connect to the universe I searched for so long since I was a kid. From being obsessed with the TV series *In Search Of* . . . to discovering Carl Sagan's world, space always fascinated me. Meeting Leland allowed me to connect even deeper with the cosmos on so many levels. In 2009 Leland took one of my BBC tees to space as well as a piece of music we worked on together, which as you can imagine meant a lot to me. He also introduced me to Katherine Johnson and has worked closely with my mother on my charity to help further educate kids from my hometown. I'm happy to be able to call Leland a friend and looking forward to many more years of learning with him. And now he will take you, too, on a journey." —Pharrell Williams

CHASING
SPACE

Amistad

An Imprint of HarperCollins*Publishers*

CHASING SPACE

AN ASTRONAUT'S STORY OF GRIT, GRACE, AND SECOND CHANCES

LELAND MELVIN

FIRST AMISTAD PAPERBACK EDITION PUBLISHED 2018.

Frontispiece: © Vadim Sadovksi/Shutterstock, Inc.

Designed by Paula Russell Szafranski

Library of Congress Cataloging-in-Publication Data has been applied for.

ISBN 978-0-06-249673-7 (pbk.)

23 24 25 26 27 LBC 9 8 7 6 5

This book is dedicated to my family: My father who exemplified Grit; my mother who exudes her namesake, Grace; my sister who shows me Love, and my niece, Second Chances. And finally, my great-niece who inspires my childlike wonder, infused with hope, and optimism.

Thank you, Mom, Dad, Cat, Britt, and C.

Love, Leland

CONTENTS

CHASING SPACE

1

I Really Can't Hear

On my first day of spacewalk training in NASA's Neutral Buoyancy Laboratory at the Johnson Space Center, the future loomed in front of me as bright as the heavens I had hoped to reach. I wanted to master the fundamentals and, like all astronauts, I knew that spacewalk proficiency was the quickest path to that coveted first flight assignment.

The sooner I could fly, the more time I would have to spend off-planet getting my orbital shift on. There are almost 7.5 billion people living on Earth. Only 555 of them have ever flown in space. Of that number, 362 were American astronauts. It's a very, very exclusive club, and as I prepared for the day's trials, I was determined to join it.

I began my descent to the bottom of the cavernous pool that

houses a submerged space shuttle and International Space Station to practice installing and attaching hardware with the goal of one day doing similar work on the real $150 billion space station orbiting high above the planet. I had completed many scuba runs in the pool before, but training for extravehicular activity (EVA) in the suit is more like walking in the weightless environment of space. It isn't easy. The suit alone made me look like a cross between the Pillsbury Doughboy and the Michelin Man.

At a depth of ten feet, I noticed the little block of Styrofoam, called the Valsalva device, was missing. The device is usually attached at nose level inside the helmet to help a diver clear his or her ears by simply pressing the nose against it. Just a week earlier, I made sure to tell the technician that I needed it.

Through the helmet headset, I told test director Greg Sims that the device was missing. I enjoyed working with Greg. We met a few months earlier in Huntsville, Texas, when he administered the qualification dives for my scuba certification. I should have stopped the dive to get a new device, but I wanted to go on with my training. So, I convinced myself—and Greg—that I could continue without any problems. "Okay," he said. "Just don't hurt yourself."

It didn't take long before my ears felt full, like the pressure a passenger feels when a plane shifts in flight—only much worse. Desperate to clear my ears, I thrust my head forward to position my nose on the helmet neck ring, but it was too far out of reach. My doctors would later tell me that either my background as a professional athlete or my high tolerance for pain prevented me from realizing the true severity of my situation. Maybe. But, on April 3, 2001, I simply told myself that my ears would clear.

I knew something was wrong at twenty feet. "I can't hear you," I told Greg. "Can you turn the volume up?" I strained to hear a response, but all I got was static. I figured something was wrong with either the headset or the cable running from my suit to the pool deck as one kink in the cable could shut down the sound.

"Turn up the volume," I shouted into the microphone, wondering why the men on the other end of the line were ignoring me.

Greg and his team began to raise me slowly up and down in the pool, thinking it might clear my ears in much the same way increasing altitude can do on an airplane. I still couldn't hear Greg. Seconds later, I heard an irritating sound from the headset, followed by faint crackles and an obscured voice. Somebody was trying to tell me something, but I couldn't make out the words. At that point, the training ended. As I floated toward the surface, I saw Danny Olivas, a fellow trainee, at the bottom of the pool ready to demonstrate his EVA skills. I know he was thinking, *What the hell is going on with Leland?* and probably wondering if my problem would delay his training.

Dr. Richard McCluskey, a NASA flight surgeon who was assisting with the exercise, was waiting for me. The technician who helped me out of my helmet was the same guy in charge of my "suit-fit check," the test run of all the equipment I would use the day of my underwater training. He had asked me if I needed a Valsalva device, and multiple scuba dives in the Red Sea had taught me that I could only clear my ears by blocking my nostrils. So I had said yes.

The look on the doctor's face told me something had gone wrong. In fact, I saw concern on everyone's faces. They were all mouthing words without sound, and I wondered why they didn't just speak up. When Dr. McCluskey touched my right ear, I noticed the blood on his finger and began to feel it running down the side of my face. He quickly examined my ears and observed retraction of my tympanic membrane—the eardrum. Suspecting the problem occurred from the underwater change in pressure, he administered a nasal spray and politzerization, a medical procedure that involves inflating the middle ear by blowing air up the nose.

Neither effort worked. I still couldn't hear.

Dr. McCluskey then escorted me to the hypobaric altitude chamber, which simulates the experience of an airplane ascending in flight. At 10,000 feet, I didn't feel any different but when the doctor tried to inflate my middle ear a second time, my ears began to clear. I could hear a little better.

According to the medical records, I could hear speech from about five feet away. However, when the chamber brought me back to sea level, my ears became blocked again, and I was unable to hear much of anything. All I could do was sit inside the steel tube, peering through the window at the faces of the doctors and technicians, and wonder if they felt any responsibility for damaging a member of the NASA family. After my stint in the altitude chamber, the room seemed to spin, and the walls, fixtures, and faces tumbled together in a confusing whirl. I threw up when I got to the shower. Thank goodness, I made it that far.

My next stop was the Flight Medicine Clinic, which wasn't far from the Neutral Buoyancy Laboratory. The doctors there examined my ears and performed balance tests that I passed easily. Not bad for someone bleeding from his ear. Still, I couldn't hear. The doctors then had me transported to Houston

Methodist Hospital, where I was examined by Dr. Bobby Alford, a renowned ear, nose, and throat specialist. He admitted me to the hospital after a battery of hearing and balance tests revealed severe bilateral hearing loss. My treatment would consist of antibiotics, antivirals, and carbon dioxide to increase blood flow to the brain. Unfortunately, my doctors could not determine the cause of my deafness.

"Suspect cochlear process as source of hearing loss," one of my doctor's medical reports stated. "Ruptured round window possible, but unlikely. Possibilities include ruptured or damaged Reissner's membrane, patent cochlear aqueduct, vascular lesions, infections, or undetermined."

Dr. Alford thought exploratory surgery might provide clues.

Just hours before, I had been on top of the world, brimming with confidence. I was pursuing my future in space. Now, I lay in a recovery room, confused, disoriented, and uncertain. I watched as Dr. Alford scribbled notes on his yellow legal pad. The eardrums were intact, and an MRI revealed no evidence of a stroke. Yet, they couldn't find the exact cause of my problem in my right ear. As much as I wanted to discover the cause of my hearing loss, I was more concerned with the space agency's response to my situation. If something bad happens to an astronaut in training and NASA can't explain it, the astronaut won't fly.

My medical status changed, and as soon as it did, the nurses in charge of documents immediately altered my records to read "DNIF," short for Duty Not Involving Flying. This medical disqualification barred me from future training in NASA's high-performance jets, and I could forget those thoughts of living and working off-planet. One mistake had brought three years of training to a sudden, dispiriting halt. I wondered if I would ever get my time in space, or be consigned to an earth-bound existence, left to look up forlornly at the stars.

"The Right Stuff"

Astronaut culture is a world Tom Wolfe famously described as divided into those who have the "right stuff" and those who don't. Some folks are just born with it, and to prove they have it, they never admit defeat or show weakness. They certainly never end up in the hospital with sudden and profound deaf-ness because a tiny block of Styrofoam was missing from their helmet.

For decades, the "right stuff" type of astronaut typically was a military pilot somehow supernaturally graced with "the ability to go up in a hurtling piece of machinery and put his hide on the line and then have the moxie, the reflexes, the experience, the coolness to pull it back in at the last yawn-ing moment." That's how Wolfe described it. A military pilot faced a 23 percent chance of dying in an aircraft accident and a greater than fifty-fifty chance of someday having to make a perilous ejection out of the jet and come down in a parachute. Yet, they still fly.

I never saw myself as one of those guys. They were the decorated officers who fought in Desert Storm and Iraqi Freedom. They talked about angle of attack and dog-fighting, spoke of days when their butts were on the line trying to save troops or keep the bad guys from hurting the United States. They had call signs, like "Dex," "Hock," "Ray J," and "Zambo." They sacrificed so much to protect this country.

I came to the space program as a civilian, not a military

test pilot. Until three years ago, I had been a research scientist, working to develop optical fiber sensors at NASA Langley in Hampton, Virginia. Before that, I was completing my master's degree with hopes of starting a lucrative career in the chemical industry. I was recruited to join NASA at a time when the space program was trying to improve the diversity of its workforce. I never aspired to a life off the ground, much less to become an astronaut. I knew that whatever path of life I was on, I wanted to do my best. But this stuff Tom Wolfe described? How could I have the "right stuff"?

The nation's space program has made progress from its days when the staff and leadership of the space agency were largely white and male. We certainly have come a long way since May 25, 1961, when President John F. Kennedy stood before Congress and announced that the United States should commit to "landing a man on the Moon and returning him safely to Earth" within the decade.

However, during that time, America's race relations were in turmoil. Dozens of Freedom Riders had put their lives on the line when they boarded interstate buses to travel across the American South in an inspired mission to end racial segregation. On Kennedy's orders, a reluctant NASA selected Ed Dwight as the first African American to join the space program. "Kennedy picked me out of the turnip patch and plopped me in the middle of this," Dwight told *People* magazine years later. "The Air Force and NASA felt someone was trying to cram a nigger down their throats."

Ed belonged in NASA. Born in the Kansas City, Kansas, area, he was an avid reader and enjoyed the arts. In 1953, he joined the U.S. Air Force, and within three years he became a test pilot. He rose to the rank of captain and earned a bachelor's degree in Aeronautical Engineering from Arizona State University. In 1961, the Kennedy administration selected Ed to be the first African American astronaut trainee. The move brought favorable media coverage, particularly among the black press.

Enduring criticisms from fellow astronaut trainees and demeaning incidents at NASA, like his assignment as a liaison to a test pilot facility in Germany that didn't exist, took their toll. Dwight told *Ebony* magazine that Kennedy's death in 1963 prompted him to leave the space program, but that seemed like only part of the reason. Another twenty years would pass before the first African American would fly in space.

• • •

I was well aware just how quickly an astronaut could be disqualified from flying. The annual physical with the NASA flight surgeon is something to be feared. Every astronaut knows you can walk into the doctor's office a test pilot and walk out a desk jockey. Physicals by flight surgeons are so dreaded that pilots often go to great lengths to have medical procedures done off-base, sometimes even in secret.

Alan Shepard knew this as well as anyone. In the late 1950s, Shepard had been one of the original seven astronauts recruited by NASA for Project Mercury. These handpicked former military test pilots, "fighter jocks" who would become known as the Mercury Seven, were fiercely competitive and entirely focused on beating the former Soviet Union into space. The race was on, and nothing was going to stop them. It was widely known that the blue-eyed Shepard was the most competitive and cocky of the bunch.

What happened next was a setback for the American space program. On April 12, 1961, Russian astronaut Yuri Gagarin climbed into his Vostok spacecraft and became the first human to journey into space and orbit the Earth. He beat the United States' *Freedom 7* mission by a scant three weeks.

Still, the May 5 *Freedom 7* spaceflight made Shepard a national hero. He was honored with parades on both coasts and received NASA's Distinguished Service Medal from President John F. Kennedy. His place in America's space program seemed secure. Or was it?

Two years after becoming the first American in space, Shepard began to have a terrible pain in his ear as he trained to command the first Gemini mission. He was dizzy and sometimes found himself unable to walk. The NASA medical team soon diagnosed him with Ménière's disease, a then incurable condition of the inner ear that causes nausea and dizziness, and in Shepard's case excruciating pain. Shepard tried to downplay his symptoms to no avail. In early 1964, he was grounded, disqualified from flying. He was barred from even flying a Navy jet without another pilot on board. He was assigned a management role at NASA—chief of the Astronaut Office. He continued to suffer and all but gave up hope of ever flying in space again.

Fortunately for Shepard, his friend and fellow astronaut Tom Stafford told him about an ear, nose, and throat specialist in Los Angeles who was pioneering a new procedure to correct Ménière's disease. The process called for implanting a small tube in the inner ear to drain the fluid. It was against NASA protocol to seek outside medical care, but Shepard checked himself into a Los Angeles hotel under a pseudonym to keep the operation away from both NASA and the media. The surgery was a success. Shepard's symptoms were gone.

It took Shepard two years to convince NASA to let him fly

again. In 1971, at age forty-seven, Shepard commanded the *Apollo 14* and became the fifth and oldest astronaut to walk on the moon. During that mission, he made more history by managing to hit two golf balls on the lunar surface.

Many years later, Chris Hadfield would come close to a medical disqualification during preparations for his 2012 trip to the International Space Station. Hadfield suddenly experienced a painful intestinal infection stemming from an appendectomy he'd had as a child. Only by working outside the NASA network was he able to get the surgery he needed. Luckily for Hadfield, the cost to the astronaut program of his not flying was greater than his medical risk, so after months of analysis he was allowed to take flight. The Corps is full of stories that ended in disappointment for the astronaut. I didn't want mine to be one of them.

• • •

Shortly after I was admitted to Houston Methodist Hospital, I met a nurse named Nancy who began to run a battery of hearing tests that would soon become a daily routine. Her teeth flashed as she smiled, but her eyes told a different story. They were full of fear—fear that I might not regain my hearing and what that would mean for my life as an astronaut. At first she didn't understand that I couldn't hear at all. She kept asking me to hold up my hand every time I heard a beep in my headset. As the machine went through its paces, I never raised my hand. *Are they fooling with me again? When is the beeping going to start?* Knowing this most likely meant that my astronaut career would soon be over, Nancy turned her head and started to cry.

The day after my surgery, my sister, Cathy, and her husband arrived from Virginia, as did my close friend Mary. Mary and my parents were connected through the deep Christian faith they shared, and she had become like part of the family. In the days following the accident, somebody called her to come down. The three of them stayed at my bedside monitoring my progress and supplying me with endless doses of optimism, shared via the yellow legal pads that had become my link to the world. My mother, Gracie, was at home in Lynchburg, taking care of my father, who was in treatment for prostate cancer. I would learn later that friends, neighbors, and people in my hometown whom I didn't even know were praying for my recovery. Prayer groups were established, sermons preached. Word had spread that one of their community members was in trouble.

Soon my hospital room was flooded with NASA visitors who had heard about my predicament from rumors that spread like wildfire through the space community. My doctors ultimately decided the visitors were hampering my recovery and instructed the nurses to start turning people away. Around this time, I got a visit from NASA administrator Daniel Goldin, who had been getting updates from Dr. Jim Locke. A few months later, President George W. Bush would accept Goldin's resignation over disagreements about his leadership of the Mars program and his overall direction of the agency. During that time, NASA's workforce went from 25,000 to 18,500. But on that day, Goldin showed up to give me encouragement.

"I will never give up on you," Goldin wrote on the yellow legal pad, and he handed me a photo of the Eagle's Nest Nebula, perhaps the most iconic image showing the beauty of space. A young cluster of stars in the constellation Serpens, it derives its name from its resemblance to an eagle. The Pillars of Creation remains one of the most recognizable regions of space. The Hubble Space Telescope captured the cluster of stars and gas in a memorable photograph that features three mounds that resemble stalagmites. Estimated to be more than five million years old, these stars have inspired countless astronauts to go to space.

I was not one of them. While I had a passion for doing many things, my goals hadn't included exploring the cosmos. The universe pulled me there. I somewhat serendipitously ended up at NASA because of a tenacious recruiter. I never imagined space travel until the possibility was presented to me. Just hearing the words that I would make a good astronaut changed me. Getting to my ultimate goal of orbiting the Earth involved a process of mastering many challenges, which now included recovering as fast as I could.

• • •

One of the first things an astronaut trainee does is select a classmate to function as crew support in the event of a medical emergency such as mine, and I had chosen Garrett Reisman for the job. Garrett would eventually clock 107 days in space over one shuttle mission and one long-duration flight on the International Space Station. But for the time being, he was my link to NASA and the rest of the world. The moment I was pulled from the pool, Garrett sprang into action. He kept in touch with my folks in Lynchburg, stood at the ready to make sure I understood the tests being administered, and brought me clothes and supplies from my house.

Before long an inquiry was in full swing, led by astronaut and flight surgeon David Brown, who was now in charge of my "mishap" investigation. Dr. Jon Clark was the liaison for the Flight Medicine Clinic. NASA protocol ensures at least one astronaut takes part when there are any problems involving another astronaut, and I was happy the two of them were assigned to my case. Both men were considered examples of NASA's finest flight surgeons. From my hospital bed that day I would never have believed that only two years later, on February 1, 2003, David and six other friends would lose their lives when space shuttle *Columbia* broke apart over Texas.

Jon was particularly dedicated to seeing me fly again and had researched new procedures for restoring sudden hearing loss. One promising experiment involved injecting a steroid into the middle ear. Jon said the Army was having success using this method to treat sudden hearing loss from acoustic trauma on the target range and battlefield, and he thought I should consider being part of the trial. But Dr. Alford would have none of it. Intent on ensuring no further harm was done to my hearing, he would not approve any experimental procedures.

Despite my doctors' reservations, I received a steady stream of visitors over the next few weeks as I endured countless procedures and prayed for a miracle. One man who stopped by was Willie McCool, the shuttle pilot who would eventually perish at the helm of the *Columbia*. I still have a drawing from one of his sons, Cameron, that shows me flying out of the hospital with praying hands. To this day that picture serves as a reminder of my old friend.

Many of my visitors spoke optimistically, but I had become a skilled reader of faces. I could tell that most of my fellow astronauts had little hope that I would ever fly in space.

The most notable example was a decorated Navy fighter pilot and NASA astronaut who flew on three shuttle flights and was commander on one. He was an exceptional flyer who graduated first in his class at the U.S. Naval Academy and received nearly every award possible. He knew how the agency worked. Angry that NASA had injured one of its own, he wrote his words out carefully but with brutal honesty. He told me he saw no chance I would ever fly in space. His advice? File a lawsuit against the agency and try to get paid for the time I had already invested. Then, he said, write a tell-all book about it. I told him I would think about it, but I knew I wouldn't. I was far from ready to give up.

It wasn't until much later that I learned all the rumors that circulated throughout NASA about what had happened that day at the pool. Despite the abundance of advanced degrees and even Nobel laureates who work at NASA, the agency's rumor mill isn't always accurate. Some people surmised that I had simply passed out. Woodrow Whitlow, a buddy from NASA Langley who had gone on to be the director of NASA Glenn Research Center and later associate administrator for Mission Support at NASA headquarters, was at a technology convention in Las Vegas at the time of the accident. "Right away there were rumors about what might have happened," he told me much later. He believed I had had a stroke, which was a prevalent scenario making the rounds. The craziest rumor floating around after the incident claimed I had died underwater but mysteriously came back to life when I was brought to the surface.

● ● ●

From the time I was accepted into the Astronaut Corps three years before, everybody seemed convinced that I had what it would take. I was athletic and muscular, and had a graduate degree in engineering. I was smart and good in school but not great, though when faced with certain tasks that provided just the right kind of challenge, I mastered them. I was a quick study, agile, and unflappable—some would say fearless. But would that be enough? Now I was laid up in a hospital bed unable to hear a bomb drop.

I remember one evening watching *Good Will Hunting*. Matt Damon plays the title character, a brilliant night janitor at the Massachusetts Institute of Technology who solves a nearly impossible equation on a blackboard in the math department. The soundtrack is a melodically beautiful instrumental piece by Danny Elfman that to this day I associate with the gratifying sense of the janitor's accomplishment. The scene has inspired me in the past, but now I couldn't hear it. I couldn't hear the music. I slammed the laptop shut and started to cry. What upset me the most was the possibility that I might never again hear Prince's "Purple Rain," Aretha Franklin's "Respect," or Vivaldi's "Four Seasons." Space travel was an anticipated thrill, but music had always been an important part of my life. I didn't want to imagine living without it.

Still, this was not the first challenge I had encountered in my journey to space. In the thirty-four years before I was selected to become an astronaut, I experienced my share of setbacks and failure. Each time I stumbled, I got back up and tried again. I never gave up, and through each challenge I became better equipped for the next. So, with all the faith and prayers I could muster, I remained optimistic that I would someday hear a concerto while floating in space.

2

Vision, Grit, and Lynchburg

grew up wanting to be Arthur Ashe. His triumphs on the tennis court were the stuff of sports legend. The first black player ever selected for the U.S. Davis Cup team (he eventually became captain), the first black man to win the U.S. Open, and the first to win Wimbledon, he paved the way for subsequent grand slam champions such as Yannick Noah and the Williams sisters. But he wasn't just one of the greatest athletes of his era. He was also an author, scholar, and activist. On and off the court, Ashe was celebrated for his brilliance, toughness, and strength of character.

The trajectory of Ashe's budding reputation as a tennis player got a considerable boost in the summer of 1953, when he left his home in Richmond to stay in Lynchburg, Virginia. He moved into the Pierce Street home of Dr. Robert Walter

"Whirlwind" Johnson, a renowned tennis coach and also the first African American physician to work at Lynchburg General Hospital, where I was born. Five years before, Dr. Johnson had coached the legendary Althea Gibson on Pierce Street after hearing about her victories at the American Tennis Association championships.

Ashe practiced with Johnson every summer until 1960. His mentor taught him much more than the fundamentals of the game, infusing in him the sportsmanship, composure, and grace that would earn Ashe the respect and admiration of millions of tennis fans, including my dad. Ashe was a gentleman and a champion, and I wanted to be just like him.

My family lived on Pierce Street for a few years in the late 1960s, having moved there from an apartment owned by Dr. Johnson on Fifth Street. I never got to meet Arthur Ashe when I was a kid growing up in the neighborhood. By that time, Ashe was far from Lynchburg and well on his way to becoming a tennis icon. In fact, it would be many decades later when I realized the powerful legacy of Pierce Street and its role not just in creating great tennis players, but also in shaping some of the most influential African American voices of the last century.

Just a few blocks from Dr. Johnson's was the home of Anne Spencer. She was a poet and a civil rights activist at the vanguard of the Harlem Renaissance exploding up north in the early part of the twentieth century. Over the years, George Washington Carver, W. E. B. DuBois, Thurgood Marshall, Dr. Martin Luther King Jr., and other political and literary personalities all visited her Pierce Street home to help further the cause. The Lynchburg chapter of the NAACP was founded in her living room.

Anne raised three children in that house—Bethel, Alroy, and Chauncey, who grew up to become a black pilot who helped create opportunities for generations of black airmen. In 1939, he and his buddy, Dale White, climbed into an aging Lincoln-Page biplane they dubbed *Old Faithful* and made a ten-city tour that ended in Washington, DC, gaining the attention of the media. Their newsworthy encounter with then senator Harry S. Truman on the steps of the Capitol prompted Truman to convince Congress to permit the training of black civilian airmen. He secured $3 million in public funds, the seed investment for the Tuskegee Airmen and the first step in opening aviation to African Americans.

It wasn't until I returned home after my shuttle missions that I learned how meaningful Chauncey Spencer and the other Pierce Street legends were to my life. Their accomplishments in sports, the arts, medicine, and aviation created a legacy that cleared a path for my success and that of many others. They changed the course of history by opening the skies to anyone with the determination to explore the unknown and advance our knowledge of the universe. In so many ways, they helped shape my future as a space explorer.

But what was it about Dr. Johnson, Arthur Ashe, and Chauncey Spencer and his mother, Anne, that led them to persevere in the face of such long odds? And why do some people reach their potential despite the obstacles in their path while others just as talented do not?

One school of thought maintains that success in the face of staggering odds requires a potent combination of self-control and "grit." University of Pennsylvania psychologist Angela Duckworth describes it as the ability to sustain focus on a long-term goal, to keep at it. "Woody Allen once quipped that

80 percent of success in life is just showing up," Dr. Duckworth said. "Well, it looks like grit is one thing that determines who shows up." I would begin to hear a lot about this essential quality when I became head of NASA Education, but I was familiar with it before I knew it had a name. It's the ability to keep working, day after day, toward some goal way out on the horizon. Summoning my own grit helped me rise from my Houston hospital bed, when it seemed I would never fly again, to see the vast reaches of the cosmos via the International Space Station. The roots of my determination extend all the way back to Pierce Street.

From Pierce Street to Hilltop Drive

Before I started elementary school, my parents decided to leave Pierce Street, moving across town to buy their first home in a community filled with educators, like them. For Deems and Gracie Melvin, the move to Hilltop Drive in Lynchburg's Fort Hill neighborhood was no surprise. My parents were "strivers" in every sense of the word. They believed in discipline, hard work, and the Golden Rule. They treated people as they wanted to be treated. They lent a helping hand to their friends and neighbors and, as Christians, they understood the power of prayer and knew that it could change things for the good.

Our family moved into a brick Cape Cod house across from Fire Station No. 3. The houses in the neighborhood had yards with lots of space, and our neighbors had each others' backs. Being new to the area, my sister and I needed that support. At the time, school officials in Lynchburg were under pressure to integrate the public schools, and busing kids in and out of their respective neighborhoods to achieve racial parity was the controversial answer to decades of Jim Crow laws that locked blacks and whites into racially segregated schools. Fortunately, Cathy and I were able to stay at our local school, and our new friends and neighbors eventually felt like family.

Many years passed before I understood what impact those neighbors had on my life. No matter where I went I was seldom far from someone who would discipline me if I stepped out of line. My parents were well known in the neighborhood and respected throughout town. As a result my sister and I never got away with anything. The parents shared an unspoken agreement. They understood they all had a stake in the successes and failures of the community's most valuable asset: its kids.

When I was in Mrs. Martin's fourth-grade class, Brandon Miller and I were roughhousing in the classroom and knocked over a desk. Mrs. Martin was furious and dragged us to the principal's office by our ears. She turned us over to Mrs. Carwile who gave both of us the kind of paddling I still remember to this day. But I knew that wasn't the end of it. Mrs. Jones, the mother of my best friend, Butch, and also a teacher in the Lynchburg City Schools, had heard about it before I even walked through her door that afternoon. "Leland, come over here," she said calmly from the kitchen. When I got there she sat me down and started in. *"What in the world were you thinking, behaving like that in school? You know better."* Mrs. Jones was a second mother to me and I hated to disappoint her. The scolding I got from her that day hurt more than Mrs. Carwile's spanking.

But I knew the worst was yet to come—my father's reaction. That evening I happened to answer the phone when Mrs. Carwile called to talk to my folks. Before my dad picked up the receiver, I made sure I was up in my room, far from the conversation. A few minutes later I heard his footsteps on the stairs. I imagined the size of the switch in his hand. I now realize how many people back then felt they had a stake in my future. Dad handed out the heavy discipline that day. However, in our house, my mother and father were a unified team—if we were in trouble with one, we were in trouble with both.

My mother, Gracie, grew up on a farm in Halifax County, Virginia. As the oldest child, she had her share of chores, from feeding the chickens to milking the cows—doing whatever needed to be done. Her parents simply expected that she work hard to help the family, and she embodied that as my mother with her own family.

My mother's faith fueled her convictions. She was deeply involved in our church, Jackson Street United Methodist, and she made sure my sister, Cathy, and I were there every Sunday. At home, she was the supreme nurturer, sewing our clothes and cooking. In the spring and summer we grew our own vegetables, and in the fall we canned fruit in our kitchen. She made clothes for me and my sister that rivaled anything purchased off the rack. By example, she instilled in us the value not only of doing a job but of doing it well.

Learning was also important to my mother. She would read to my sister and me every night. Books like *The Little Engine That Could* and *Curious George* were transformative and unlocked possibilities in my young mind. Curiosity and the "I think I can, I think I can" refrain from *The Little Engine* were instilled in me early and helped inspire my desire to achieve.

My mother taught home economics at Linkhorne Middle School, the same place where my father taught language arts. Her students learned to sew and cook and to adopt the manners and etiquette that would prepare them for life in the South in the 1970s. For many of her female students, my mother was the only adult they trusted. Some of them were being raised by single mothers who were barely adults themselves.

I don't recall my parents ever telling us to study or pressuring us to get good grades; they simply expected us to be good and to work hard. Our parents had their lives, and Cathy and I had ours. Everyone was expected to carry their own load. They didn't hover over us or badger us to make sure we got our homework in or studied for a big test. I didn't always get A's, but I always worked hard because that was the unspoken expectation.

When I was about eight years old, my parents bought me a chemistry set—one of those age-inappropriate models that had the potential to do some real damage. The kind that was soon taken off the market. One afternoon I had mixed a concoction that burned a hole in the carpet and sent smoke throughout the house. My mother rushed into the room expecting to find me injured. Instead she found me sitting on the couch with a dumb grin on my face. Somehow, I had used the power of those clear liquids to unleash fire and brimstone in my house. How could that much power be contained in such a small amount of liquid? I had to find out.

I had heard of people becoming scientists. My mom experimented with different ingredients every time she prepared a meal, but I never made the connection to a possible career— until that moment. Thanks to the explosive power of that concoction, my young mind was awash in imagination. I'm sure I got a spanking, and while I understand attitudes toward corporal punishment have changed since my childhood, I understood my mother had only my best interests in mind.

Unfortunately, some people weren't raised with those "Golden Rules" values my sister and I learned in our home. There were two boys in particular whom I will never forget. They were older than me. I thought they were good people. I was five years old, a naive, trusting youngster who thought he had found friends. So, when they coaxed me into their garage one afternoon I went willingly. I didn't know their plans were to take advantage of me—sexually.

I try to forget, to sweep the incident into the deep recesses of my mind. There's a level of guilt and shame. Feeling like a sucker for allowing it to happen and regret for not fighting back hard enough to stop it. As much as I'd like to forget, there are always little reminders that bring those painful memories rushing back—a certain glance, white flaky paint, the smells of gasoline and fresh-cut grass.

I never told my parents, especially my dad. He would have killed those boys. He would have ruined his own life to avenge mine. I had friends without fathers, and I was not willing to risk losing mine. So, I stayed quiet. My form of self-preservation was to act as if the incident didn't happen. My family and I continued to live in the neighborhood, and I would see both boys often enough until I left for college. One of them still lives in Lynchburg, and a few years back I ran into him. We talked, but the afternoon in the garage never came up. Maybe he had forgotten about the encounter, or perhaps, like me, he's repressed it to overcome any lingering feelings of guilt, remorse, and shame.

Many people tend to suppress bad things that happen in childhood. However, the incident did force me to take a harder look at my identity and my place in the world. I was a black nerdy kid who liked athletics but was still an inquisitive introvert who was a little goofy. What was I chasing and would I always be a victim of abuse? Looking back on it, I believe I survived that horrible ordeal because that's what happens when you're fortunate enough to have unwavering faith and the unconditional love of family and community. If nothing else, getting beyond the incident enabled me to continue to grow, believe, and dream, as everyone must do to flourish in life.

My Father's Vision

Among the many lessons I received from my dad, I learned the importance of visualizing a goal and resolutely pursuing it until I achieved it. For example, one scorching hot morning when I was about eleven, my dad drove up to our house in a Merita bread truck and parked it in our driveway. He had worked extra jobs, but I had never seen him deliver bread. I knew he performed as a drummer in a band because I spent summers as his roadie, busing his speakers around town even though I was too young to be in the clubs and watch the shows. The six-foot-tall Peavey speakers towered over my small middle-school frame.

A bread truck? I asked him what he planned to do with it. "It's going to be our camper," he said. But I couldn't see it. I argued with him, telling him it was nothing but a bread truck without the bread and it would be a terrible place to sleep. It was cavernous inside, with only one seat for the driver and stacks of metal racks in the back. It even smelled like freshly baked bread. But within a few weeks, I began to understand his vision and believe in its potential.

My dad was born in 1930 and grew up in Roseboro, North

Carolina, where he learned the value of hard work. Necessity was truly the mother of invention for my father. As a kid, he *made* his first bicycle out of parts and scraps from other people's used bicycles. As a teenager, he cut trees in the woods and then worked at the local plywood mill, processing the wood. Employees at the mill didn't have power saws back then. My dad used a crosscut saw, which he said helped him to become "very strong" and created "a good work ethic." Education was important, too. He graduated from high school and went to St. Paul's College in Lawrenceville, Virginia, where he played football and ran track. He initially wanted to major in interior design, but after a four-year stint in the U.S. Air Force, he changed his major to elementary education. He went on to teach in public schools for thirty years.

In our family, whatever projects my father undertook became my projects. Such was the case with the bread truck. We installed two bunks for my sister and me and a pull-out couch for my parents—all bolted to the floor of the truck. A camping stove and table made up the kitchen. As campers go, it wasn't elegant, but it was functional and had everything we needed to go on our habitual summer outings.

Through the bread truck conversion and other projects I learned that with vision you could create something great. My father had a similar vision for our family that he orchestrated every day—at home, at school, and throughout Lynchburg. As a childhood friend recently told me, my parents weren't just committed to building their family—they were building the whole community. I would often see my dad sitting with older boys, offering guidance or just listening. Everyone knew "Mr. Melvin," and everyone my dad spoke with felt respected.

Mr. Melvin's House

My dad's compassion and civility extended even to neighborhood teenagers with difficult home lives. They were warmly received at our place. "There were houses in the area where parents wouldn't let me in," my friend Stan Hull told me. Stan and I had known each other since elementary school and played football together at Heritage High. "I came from a different part of town. But I knew I was always welcome in Mr. Melvin's house. He saw promise in everyone." Ralph Wilson, another childhood buddy who still goes by the nickname "Chopper," expressed a similar reverence for my dad. "You always felt like he was looking out for you," he said.

One night one of my dad's acquaintances came to our door while my sister and I were watching TV. He was homeless and in some sort of trouble, we learned later, and had come to my father for help. My sister and I flashed each other a skeptical look that failed to escape my father's notice. Dad invited the man in and did what he could to help. The minute he was gone we got a harsh lecture about the importance of showing respect to everyone, regardless of their station in life.

Robert Flood was another local man who knew my dad. As a kid, he lived with his grandmother, who doted on him and made him feel loved unconditionally, even when he started getting in trouble in middle school. He was exceptionally talented on the gridiron and the basketball court, but he was disruptive at school. He could have played basketball in college but he was drawn to the "fast" crowd, eventually getting kicked out of high school. When his grandmother died, there was no one. Except for my father. "Mr. Melvin never abandoned me," Robert said to me not long ago. When Robert was arrested and jailed on a felony, my father was there to help him, even though he had a young family at home. On a recent afternoon, Robert tearfully described the impact my father had on him. "He took the time," Robert said. "He listened and he treated me like I was his son."

Robert's life took many turns, with multiple arrests, bouts of drug addiction, and a history of gambling and petty crime. On June 16, 1989, however, he walked out of a treatment center, ready to start fresh, and my dad was there. "It was like he was just waiting for me," Robert said. "He was so happy that I had changed my life. I had somebody who really cared about me." Robert eventually landed a job but needed transportation. My father found him a car—a 1976 Dodge, which he offered to Robert on the condition that he stay employed and out of trouble. He did. He graduated from college and started working on a master's degree. Years later, when Robert spoke at my dad's funeral, he said he felt like he'd lost his own father. Robert wasn't the only one who felt this way.

My father had a passion for the gospel, and he believed he was put on this Earth to serve others. When I was in college, he acquired a dilapidated 1954 Ford truck that looked like a cross between a moving van and an RV. In much the same way he converted that bread truck into a family camper when I was a kid, he turned that Ford into a mobile bandstand and pulpit. He and his buddies would park the truck next to the basketball court on Fourth and Federal Streets, set up a few dozen chairs, cook up some hot dogs and hamburgers, and play gospel music. My father played keyboard, trumpet, and drums. He sang, too. My father had a smooth voice that I loved to listen to. Boys from the neighborhood and around town would come to shoot hoops and listen to my father preach the gospel—before they were fed.

My mother believed in God as devoutly as my father. Although my mother's manner was more reserved than my dad's, her impact was just as powerful. Their vision for my life and for my sister Cathy's life came from their limitless Christian faith. I would never have made it to space without calling on deep reserves of spirit derived from their loving insights.

Becoming an Athlete

For some kids in Lynchburg who didn't grow up in households like mine, sports provided a semblance of a family, and coaches gave them the closest thing they had to a father. My coaches at Heritage High took on such tasks with grace and dignity. Coaches Mark Storm and Jim Green focused on building character as well as athletes. Few did so with as much commitment and enthusiasm as Rufus Knight.

Coach Knight joined the coaching staff of Heritage High in 1976 and worked as a teacher, coach, and offensive coordinator in the football program. Even after Coach Knight retired, he continued to help coach the track team. A deeply religious man who had been a U.S. Army Ranger, he held killer workouts. But he never made us do any exercises he couldn't do himself. He is still in perfect physical condition to this day.

Like my father, Coach Knight lived his faith. If you played on his team he took care of you. Some of the players had no way to get home at night after a game or a late practice, and Coach Knight would drive all over town dropping students off at their houses. Lynchburg was still largely segregated, and the school district bused kids from one neighborhood to another to integrate the schools. Some of the boys at Heritage lived in the poorest parts of town. Years later, Coach Knight told me how he would wait in his truck as each boy went inside. He didn't leave until the family turned on the porch light to signal that there was someone home. On many nights, Coach would go to the door or walk around to the back of the house to make sure the boys were safe.

Coach Knight's caution notwithstanding, I don't recall growing up with much crime or violence in Lynchburg. Heritage High was a large, newly integrated public high school, but we had little friction. Everybody seemed to get along pretty well, as surprising as that seems to me now.

Once football season ended, I moved right into basketball season and then tennis, with little time for anything else. I had a posse of five friends from childhood. We called ourselves "The Big Blue Crew," after the second-team basketball squad at the University of North Carolina. Each of us had a nickname that stuck. Phil Scott was "Silky Blue." Kip Hawkings had an easy nickname, "Gus." Ernest Penn was known as "Fufu," and we called Bryant Anderson "Boogie Bear." My nickname? "Lil D" in honor of my dad, of course. We went through many grades together, played a lot of basketball, and learned a good deal about life. Our friendship taught me the value of trust and teamwork. I owe a lot of my success to my guys in The Big Blue Crew.

My development as an athlete had also taken place outside school, on sandlots and playgrounds. One year my dad petitioned the Lynchburg Recreation Department to build a public basketball court and park up the hill from our house. He had a gift for persuading people to do things for the community and within a few months the park was built. The Fort Avenue Park became a basketball mecca on Sundays after church. Cars lined up and down Randolph Lane and even the firemen from Station 3 would abandon their post to observe the intense play and occasional fight. The small park was filled with so many ballers that if you lost your first game you wouldn't play for the rest of the day.

I first played there as a young kid competing against grown men, many of them high school standouts who didn't make it to college due to poor grades, brushes with the law, and drugs. Some had earned scholarships but for whatever reason couldn't adjust to college life and after a semester or two away from home found themselves back at the neighborhood basketball court trash-talking and showing their skills. While I waited between games I learned about life on the street and the undercurrent of what was going on outside of the idyllic image of Lynchburg. I learned from talking with the older guys about neighborhoods across town plagued by drugs and poverty. Those conversations made me realize that I had a naive image of Lynchburg, thanks to my parents shielding Cathy and me from negative influences. I am grateful for my parents' efforts, but life lessons can come from anywhere, including the basketball court in between games.

The court also gave me "cred," especially when there were fifty-plus people watching and ten teams of five waiting to get their game on. Never did I think I would someday be looking down at my old neighborhood from the International Space Station, orbiting the planet with roughly 240 miles separating me from the basketball court and my perch in space. From up there, where the world is really small and there are no boundaries or borders, I imagined the court filling up after church.

I remembered long days hustling across the asphalt, my eyes on the hoop.

Basketball wasn't my only passion back in high school. I had a girlfriend during my senior year, a young lady I had known for some time. We started off as friends and then we became study partners. At Heritage High School, we grew closer. On the night of our graduation, she and I drove to a parking lot behind an abandoned strip mall where we could be alone. We sat in the darkness, making out in the 1964 red Peugeot my dad bought me in the summer of my sophomore year. I had recently accepted a football scholarship at the University of Richmond. She was headed to the University of Virginia.

Suddenly, out of nowhere, a state trooper was standing at the side of the car shining his long-handled flashlight down on us. He had crept up behind my car in his cruiser with his headlights off. *We're screwed*, I thought. My girlfriend and I exchanged worried glances. We both knew there were plenty of cops around Lynchburg who wouldn't appreciate our being together as we were not of the same race.

"Young lady, please get out of the car and join me in the patrol car," the trooper said. Nervously, she opened the door and got out. She then walked over to the patrol car and sat next to the officer in the front of his cruiser.

"Who is that you're with? What are y'all doing?" the trooper demanded. "Young lady, that man was trying to rape you, wasn't he? If you don't tell me, I'm going to take you both down to the jail where your parents will have to pick you up." I suspect that tactic worked more often than not, but luckily for me, my girlfriend didn't take the bait.

"That's my boyfriend. He's a good guy," she said. "We just

graduated. We're about to go to college. He's got a full scholarship."

My girlfriend later told me that the trooper had tried to pressure her to claim that I had assaulted her and that she was there against her will. Fortunately, she stood her ground and stuck to the truth. The trooper then got out and brought me into the patrol car too. He had me sit in the backseat. He asked me what I had been doing in my car. In that moment, surrounded by the flashing red and blue lights in the backseat of a police car, I became vividly aware of what was at stake. Going to Lynchburg City Jail on a trumped-up charge could be a game changer. I could kiss my scholarship goodbye and perhaps a college education, too. The state trooper had the power to change the course of my entire life—for the worse— just because he felt like it.

For too many young black men, an encounter like this with the police ends badly. It was the norm for some guys I had grown up with in Lynchburg. Maybe their girlfriends panicked and buckled under the intimidation. Or maybe they mouthed off at the officer and things spiraled out of control. Once a young black man falls into the criminal justice system, he often keeps falling. The trooper seemed determined to add me to that number, but for some reason he changed his mind. "I could have taken you to jail," he said, "but I'm doing you both a favor tonight. Get outta here. Don't park here again."

My First Second Chance

Graduation was the cap on a string of victories and advances that had begun to unfold earlier that school year. At our homecoming game the previous fall, we had been down by a touchdown with only minutes remaining in the game against the Rustburg Red Devils. Several college football scouts were in the stands looking for exceptional talent in a game we were favored to win.

There were only minutes left to play in the game. I had played wide receiver on the varsity squad since my sophomore year. I flanked out wide on the 50-yard line and quickly looked at the defender, then back at the ball. Many alumni had returned from college eager for a win to start the homecoming festivities. The roar of the crowd made it impossible to hear the snap count that would set everything into motion.

I saw the snap of the football and started down the field. The defender tried to jam me off the line of scrimmage but I countered and rushed past him. In my periphery, I saw the fans on their feet as Stan Hull, our quarterback and my buddy since elementary school, launched a tight, perfect spiral that was on a trajectory to meet me in the end zone.

At that moment, though, I was adjusting my speed to ensure a harmonious meeting of my hands and the ball for the win. Our fans were on their feet screaming. I think some were already celebrating. Then suddenly, silence. I had done the unthinkable. I dropped the pass in the end zone. As I looked to the stands I saw people shaking their heads in disbelief. Morgan Hout, a scout from the University of Richmond, had come to see a couple of my teammates, Leonard Dempsey and Daryl Parham. When I dropped the pass, he turned and headed for the exit.

I expected to be benched for the rest of the game, but that's not what happened. Coach Green grabbed me by the facemask, looked me in the eye, and said, "Get back out there, run the same play, and catch the ball."

The coaches wanted to give me another chance. I was willing to let that dropped pass define my high school career but they were not having it. "You had worked so hard," Coach Knight said to me years later. "We just believed in you." I went back in the huddle, delivering instructions to run the same play. My teammates weren't happy with the news. I had just let them down and now I was being given a second chance. "We all wondered," Stan recalled, "if it didn't work the first time, why would it work the second?" But, there was no time for debate about the play. We broke the huddle, lined up, and on the snap I took off for the end zone. Stan threw another perfect spiral to the same spot but this time I caught it. The crowd went wild. Hout heard the roar, turned around, and walked back toward the end zone where I was celebrating the victory with my teammates. We had won the game and that catch resulted in a full football scholarship to the University of Richmond. When I joined the team that fall, Hout was my receiver coach.

I didn't understand it then, but I had been given an opportunity that would alter the course of my life. The fact that others believed enough in me to give me a second chance—even after I failed before—inspired me to persevere against the odds and to never give up. It wouldn't be the last time I would get a second chance.

3

Second Chances

stared into a plastic cup that had just been handed to me. Inside it was a swirling mass of snuff spit, Tabasco sauce, and a bunch of other dubious substances. It was my first night at the University of Richmond and the whole team was on hand as a huge three-hundred-pound offensive tackle acted as master of ceremonies and our chief tormentor. Twirling his massive arm above our heads, he snapped the tip of a leather whip. "Drink," he commanded.

Welcome to college football, I whispered to myself.

Before long I found myself tied up to defensive back Taylor Lackey, whom I'd known for only a few hours. Together we were thrown blindfolded into the bed of Billy Cole's pickup truck. Taylor and I lay back-to-back and practically buck naked, as the truck bounced along for what seemed like hours.

It was only a day into training camp and Taylor would soon distinguish himself as an outstanding safety, a fierce hitter from rural Georgia who could knock the crap out of whatever stood in his way. But on that particular night we were both helpless as we lay there in our jockstraps, sweating in the heat.

When the truck finally stopped, the players yanked us onto the train tracks and left us as they sped away. I remember hearing a train, its horn blaring louder and louder as we lay there. It seemed like we would be crushed at any second. The train never came. We found out later that the sound of an oncoming train was actually an air horn attached to Billy's truck. The challenge now was to get back to the Robins Center on campus without getting arrested for indecent exposure. First we had to get out of our blindfolds without the use of our hands. We didn't have a clue where we were and the dark, starless sky made it difficult to navigate. My stomach was still churning from the initiation phlegm cocktail I had downed.

"Where the hell are we?" Taylor yelled. As we walked, I saw the familiar lights of the library. We hadn't gone as far as it seemed—we were on the edge of campus about a half mile from the Robins Center. Billy must have driven us around in circles. We sprinted from bush to bush, hoping not to be spotted in our jockstraps.

We finally made it back to our rooms. At practice that morning, everyone suited up and stood on the field as if nothing had happened. The relentless July sun beat down as Coach Dal Shealy laid out his strategy for reversing the team's losing streak. I couldn't hear any of it. My head was pounding too loudly.

On my second night in the dorm we were led down the

hall to the big communal bathroom where the veteran players were shaving the heads of the freshman. *Could be worse,*
I remember thinking. I didn't have a lot of hair anyway. But
Don Miller, a linebacker, didn't take it so lightly and refused
to open his door. "Go to hell," he snapped. "I'm not going to
do it!" The upperclassmen kept banging, telling him to come
out, telling him he'd regret it if he didn't, and soon dozens of
us crowded into the hallway eager to see what would happen
next.

Don didn't open the door. Instead he slid open the window
of his second-floor room and jumped. Miraculously he landed
on his feet and sprinted across the courtyard. But a few hours
later, he came creeping back into the building, where he was
ambushed and whisked into the bathroom. The boys were
waiting, and in short order Don's head was shaved down to
the skin.

In some ways, those hazing rituals helped prepare us for
the rigors of Coach Shealy's training. He was someone you
didn't want to disappoint. He had played football at Carson-
Newman College, a Baptist school in Tennessee, and then
joined the Marines, where he played for the legendary Quan-
tico Marines. Coach Shealy retired two years after I had graduated, and he went on to become president of the Fellowship
of Christian Athletes, an interdenominational Christian sports
ministry. Coach never cut us any slack. He was relentless
when it came to training, pushing us to both physical and
mental exhaustion.

To all of us on the team, Coach Shealy was the model of de-
corum, a Southern gentleman from South Carolina who considered coaching his God-given mission. He forbade swearing
and routinely led the team in prayers. He didn't fare that well

stopping his players from roughing each other up and drinking to excess after the games, especially when we lost. He sought players who had character and enough grit to become part of a team.

"My philosophy was to coach the way I like to be coached," he later told me. He described the men who had guided him as "people coaches" who wanted "to teach you to be a man, to teach you to live life and be the person you should be."

Somehow he got the best out of us.

Our Losing Season

They say that in the South some traditions refuse to die. At the University of Richmond, football was one of them. The 185-year-old school had fielded an NCAA football team for more than a century before I got there. But the year I arrived, the program was in serious trouble.

For the past six seasons, the Spiders had lost more games than they had won. Every year there was talk about killing the program, and the year before I got there the critics were louder than ever. But legend has it some powerful alumni stepped in and convinced the school the program was worth saving. They cut a deal to pay for renovations to City Stadium, the 20,000-seat facility three miles from campus where the Spiders had played since 1929. The football program was saved, at least for the time being, though having to play home games off-campus made it hard to keep students interested, especially when the team was still losing. To top it all off, the football program had just been dropped down from Division 1-A to 1-AA, an NCAA decision that devastated the coaches.

That was the scene I stepped into when I moved into Jeter Hall that fall.

With the start of classes I found myself thrown into a punishing schedule of training and practices while I tried to keep up with a full load of math and science courses. It was a good night when I managed to get five hours of sleep. My mom and dad came to nearly every game, even if it meant driving a full day or more from our home in Virginia to upstate New York or even Rhode Island. My mother worried constantly that I might get hurt. I later found out she prayed I would get a minor injury so I would be taken out of the game and avoid a worse one.

During my freshman year we lost every single game and ended the season 0–10, but the worst part was that we were demonized around campus. Complete strangers snarled at me or would shake their heads in disgust. *Loser. You suck.* Being an athlete had always been part of my identity, but that season caused a lot of us to question what we were doing there.

The University of Richmond was a small school with about 2,100 students and, it seemed to me, a lot of money and an aura of entitlement. When you're on an athletic scholarship and your team is losing, school administrators, fellow students, and even professors make you feel like you've let them down personally, like you weren't doing your job, or like they paid for an expensive product that didn't operate as advertised. Of course, there's that underlying belief that certain athletes couldn't possibly deserve to be there on their academic record alone.

A few days into the school year, I sat in a meeting with my freshman advisor, Dr. J. Ellis Bell, and about twenty other students. Dr. Bell asked for a show of hands from those who were exempt from taking certain classes because of their AP test scores. "Like Leland here," he says. "He's exempt from taking calculus." I could detect a palpable sense of surprise throughout the room. *What, him? The black football player?* I could practically hear their thoughts. Some of them even raised their eyebrows.

I was relieved when the season ended and I was able to go home for Christmas break. It had been a long and hard introduction to college—particularly college football. I had never been on a losing team with an 0-10 record. I looked forward to returning home to a family that was proud of me no matter how many games we lost. Hot chocolate, a warm fire, and the scent of pine in our house on Hilltop Drive provided familiar, much-needed comforts. Exhausted from a season of game travel, training, study hall, classes, and final exams, I was relieved to finally be all done. I just wondered about my inorganic chemistry grade.

I thought I'd done well, and I liked chemistry since I was a kid. I'd had a B in the course all semester, so I wasn't too worried. To hear the professor, Dr. William Myers, tell the story later, I wasn't his strongest chemistry student that year, but I was perhaps his most determined. I didn't get discouraged at not having the answer right off or making a mistake in a hypothesis. He saw I had the potential to become a scientist and, as a supremely gifted teacher, he looked forward to helping me achieve that.

When an envelope came from the University of Richmond I knew it contained my grades. However, there was no letter grade for my chemistry class. Instead, there was an X—rather than the B I had expected. I assumed Dr. Myers had simply not finished grading finals before the break, though that was not like him. Like most serious scientists, he was meticulous, efficient, and always graded our tests within a few days of our taking them. I would also come to learn that he was a superb writer, as competent with words as he was in a lab. Still, I wasn't worried as the knotted pine popped in the fireplace and all seemed right in Lynchburg.

Honor Council

After a few weeks at home, I started looking forward to getting back to Richmond. I was tired of listening to friends constantly ask how our football team had done. They all knew we had not won a game; they wanted to rub it in about being a loser. *Big-time college football player didn't win a game.* I drove back to Richmond from Lynchburg ready for a new start, beginning with off-season football training. I was also eager to find out my chemistry grade. I went straight to my dorm and dropped off my bags. I remember getting a call on the hall phone to come to the student commons to talk to Stephen Kneeley, the student head of the Richmond College Honor Council. I figured the council had decided to invite me to join as a new member of the council. *Wow, that's pretty cool,* I thought to myself. *And I'm only a freshman.* Sure, I was busy with football and school, but I thought this would be a good chance to get noticed for something besides the gridiron.

I entered the room and exchanged greetings with Steve. The next words I heard from him sounded like an unintelligible foreign language. It was as if he were speaking in slow motion and then time stopped momentarily as I tried to grasp what he had just said. He told me I had been accused of cheating in Dr. Myers's inorganic chemistry class. Thoughts of becoming an Honor Council member vanished as I tried to figure out what was going on. The council members looked concerned, but I knew they believed I was guilty.

The charges had been brought against me by a senior named Tom, who lived across the hall from me. He was a business major struggling in what was known as "football chemistry," the basic level class for nonscience majors. Tom needed a final exam grade of B to graduate and asked me to tutor him, even though he probably hated the idea that a football player was possibly making an A in a much more challenging chemistry course. Tom was also a member of the Honor Council. He alleged that he had heard me discussing the chemistry exam I was scheduled to take with another student. The other student happened to be his girlfriend. She was a freshman and had taken the exam earlier that day. She had walked into the room where we were celebrating Tom's getting the B he needed to graduate.

The truth was I *did* hear the girlfriend describe the test, and I didn't bother to tell her to stop. I simply didn't care, nor did I need the help since I had performed well throughout the semester. She had taken a different test anyway, it turned out. But, Tom decided to turn both of us in, and we had to stand trial.

The Honor Council was composed of a dozen or so white students, mostly male, seemingly selected for their proximity to wealth and privilege. At the time, the number of minority students at the University of Richmond amounted to 0.01 percent of the student body. No one on the Honor Council looked a thing like me, the small-town football player. The fact that the Spiders had lost every game that season didn't help my case much, and instead probably bolstered the arguments of those who wanted the school to drop the football program altogether. *They're not only losing—they're cheaters!* The council was overseen that year by Dr. Richard Mateer, a university dean and by coincidence a chemistry professor.

The trial took all of fifteen minutes. I did not receive any help with my defense; no one from the school's administration ever talked to me or explained my rights. I was alone facing a panel of students who knew nothing about me but held enormous power over my future. The panel found me guilty of cheating and suspended me from the university for a semester.

After I left the trial I walked to my dorm and sat on the steps as the tears streamed down my face. The worst part was that I had to call my dad and tell him I had been suspended from school. I figured I'd lose my scholarship.

That day could easily have been my last at the University of Richmond. But as I sat with my head buried in my hands, my roommate, Dan Fittz, came out and said I had a call on the pay phone. Dr. Myers wanted to see me. Dreading what he would say, I dragged myself to his office. I had disappointed him, and he wanted an explanation. But, when I sat down in his office his demeanor was not what I expected. The Honor Council had recommended to him that he give me a failing grade in his class, but he couldn't do it. I was not, in his words, an F student, but rather a student who had made a serious mistake.

Dr. Myers knew that my failing that class could have altered the course of my academic career in ways perhaps impossible to reverse. All semester Dr. Myers had been impressed that I had effectively balanced the rigor of his course with my schedule of daily practices and near-constant travel. He decided that, since I had worked that hard, he would fight for me.

Years later when I contacted Dr. Myers to discuss the cheating accusation, he was reluctant to talk about it, regarding it as a deeply personal episode that he'd put behind him. Yet he conceded that he had considered me guilty of the charge. Regardless of whether I used it or not, I had received advance information about the exam. The question for him, he said, had been what to do about it. He felt there was no appropriate sanction, at least not among the harsh sentences that were being proposed. I had made a grave mistake that day by not leaving the room, but he was not going to let that mistake impact the rest of my life.

As part of my punishment for poor judgment I had to take an ethics course and work in Dr. Myers's chemistry lab until I graduated. "There is something wonderful about seeing the patterns that abound in nature, and there is something more wonderful about helping others see those patterns," he once told an interviewer when asked about my time in his lab. While showing me those abundant patterns, he turned me into a scientist.

I remember that the following year I took organic chemistry, a notoriously difficult course that typically weeds out weak students. Many take it over the summer so they can focus on just it, since it's a required prerequisite for medical school. There were two options: Dean Mateer's or Dr. Stuart Clough's class. I chose Dr. Stewart Clough since I had a difficult time with Mateer's Honor Council. Years later Dr. Clough would tell me that he was initially skeptical that

I could keep up because of my demands from football. But I proved him wrong. I remember him walking through the classroom dropping graded tests on the desks of each student, facedown. When he got to me, he laid the test on my desk faceup. It had a B+ written across the top, one of the highest grades that day.

"Very good job, Leland," he said. Years later, he recalled, he had been impressed that a football player had done so well. I looked down at the grade. "I could do better," I said. The next year I was awarded the department's Merck Index prize for the Organic Chemistry Student of the Year, and it was Dr. Mateer, the dean overseeing the Honor Council, who presented me the award.

More than thirty years later, I returned to the University of Richmond for a Spiders reunion and took a tour of the new athletic facilities, including a meeting room for the football team. The room had been paid for through the generosity of the Kneeley family. Stephen Kneeley had been student head of the Honor Council in 1983 when I faced charges. He had become a peer during the intervening years and had told me about the family's gift to the school. I was surprised and touched to see a dedication plaque on the wall, and realized my journey at UR had come full circle.

Tragedy and Turnaround

My professor's decision to overrule the Honor Council's sanctions did not win me any friends in some corners of the school, but that didn't bother me. What did was that people might believe I had actually cheated on the test in hopes of getting a better grade. Once again, I felt like I had to prove people were wrong about me. I also faced another challenge: keeping my friends on the team from carrying out revenge on Tom, the senior who had turned me in. My teammates were outraged. I had always been something of an anomaly on the team, a kid from the rural South who loved science. We were a team, a losing one, but still a team. My boys had my back.

Although football season was over, training continued all year. On the first day of practice after the holidays, and on any Monday following a particularly lively weekend, the team would convene in one of the smaller gyms on campus. Strength and conditioning coach Harry Van Arsdale would close all the doors and turn up the heat, to about 105 degrees. Then we did drills until we threw up.

School didn't get easier when football season wound down. I knew that working in a lab was key to getting into graduate school and obtaining that lucrative job in private industry. My lab job didn't leave me time for much of anything else.

I spent a lot of time with my professor, and maybe one other student in the lab. Where football gear included shoulder pads, helmets, and mouth guards, our lab equipment consisted of goggles, lab coats, Erlenmeyer flasks, and glove boxes for working with carcinogenic materials. *Sports Illustrated* published a photo of me in my lab getup during my senior year, holding a beaker of dry ice while mysterious vapors encircled me. Under Dr. Myers's watchful guidance, I conducted research on amine-haloboranes from the second semester of my freshman year until graduation, looking at the inductive effects of potentially cancer-curing drugs. Working in the lab sometimes required me to miss most of football practice. Over the years, my teammates took to calling me

"Larry Lab" because coaches permitted me to show up near the end, looking pristine after everyone else's uniform bore all the dirty hallmarks of a tough scrimmage. I absorbed the nickname in good spirits while pointing out that I held up my end of the bargain: The coaches allowed it because they knew they could count on me to hold on to the passes thrown to me in closely fought games.

The summer after freshman year, I stayed in Richmond to take my first physics course. I was doing research for Dr. Myers and I worked out every chance I had, hoping perhaps that every pound of muscle would help Richmond win more games. I was determined that this season would be different, that we would gel more as a team and, with some luck, turn things around.

I couldn't have been more wrong about that fall. Among the standout players that year was a huge offensive tackle from Marietta, Georgia. He was six foot six and nearly three hundred pounds, and the coaches were talking him up as a rising star. There were rumors he was taking steroids, perhaps lots of them, but nobody was paying much attention, probably because he was playing well. But four weeks into practice and only two weeks before the Spiders' first game of the season, he drove to the home of his girlfriend in Chester, Virginia, about twenty minutes from the University of Richmond campus. It was about four in the morning, and he was furious that she wanted to break off the relationship. Using a key she had given him, he let himself into the house she shared with her mother and brother, waking them up and launching into an argument in the hallway. According to police interviews with the girlfriend's younger brother, who had been in the house as well, my teammate raised a rifle and began shooting them, point-blank, killing his girlfriend and her mother before turning the gun on himself. The brother was also shot, but the doctors managed to save him.

The tragedy stunned the quiet university community and the city of Richmond, leaving the Spiders distraught. Rumors were widespread that a dependence on steroids had warped the player's mental state far worse than anyone had imagined. I never saw steroids and was never exposed to them during my time on campus. Our coaches never pressured us to consider them. Still, questions remained. Why hadn't the team seen the signs? How much had he been using and for how long? There would be no answers. The stage was set for what would be another challenging year on the field for the Spiders.

We won more games that season, an improvement but still not enough to boost our reputation on campus. The tragedy of our teammate had taken its toll. We ended the season with a 3–8 record. We were still losers.

The Ground Shifts

The Spiders' performance on the field changed dramatically during my junior year. It was as if the ground shifted; we started to win. The mind-set of the team was transformed from one of disappointment to one of constant improvement and optimism. The starting lineup from my first two years had graduated. Coach Shealy finally had the team he had set out to build when he arrived at Richmond. His approach was to look first for character in an athlete, and then help him develop into a champion. After years spent looking for the "right stuff," he felt he might finally have it in this team.

Often on the night before a game, Coach Shealy would

come into my hotel room and sit at the foot of the bed. "Leland, close your eyes," he'd say. "You're lined up on the thirty-yard line. You're running a post corner into the end zone. You're accelerating past the defender. You're now looking at the ball. The ball is coming into your hands. You tune out the crowd. You're now in the end zone. You caught the ball. We've won the game."

Sleeping with that visual helped get me ready. Once I got on the field I had already done it. Visualizing a pattern, a defensive coverage, or move was a really important part of the headwork to prepare for the game. This type of practice would serve me well years later when I operated the fifty-eight-foot robotic arm on space shuttle *Atlantis*. I practiced my maneuvers over and over on the ground through simulation and thought about it so deeply that when it came time to perform the actual installation of the Columbus research lab onto the International Space Station, I had mentally already completed the task.

• • •

I was starting every game now, and quarterback Bob Bleier threw every pass I caught and I caught every pass he threw. We were a force. To hear Bob describe that year, "We were a nice duo. We complemented each other. We made each other famous." Halfway through the season I was neck and neck with a receiver named Jerry Rice in the number of catches per game. We were 8–4 and went to the playoffs.

By my senior year, the Spiders were the number-one-ranked team in the nation in Division 1-AA after seven games, averaging thirty-one points a game. I caught sixty-five passes, eight for touchdowns, and totaled 956 yards. It was an exciting time—we were winners and the school loved it. Dr. Clough, my organic chemistry professor, told me years later of going to a home game back then with his two young children. The three of them were cheering and yelling my name, whooping every time I caught a pass. Somebody tapped Dr. Clough on the shoulder.

"Excuse me," a woman sitting on the bleacher behind him said. "Is that your son?" It was a fun moment for Dr. Clough. "No, but I wish he were," he said. Many years later, both Dr. Clough and Dr. Myers would drive down to Cape Canaveral to witness the launch of space shuttle *Atlantis* and cheer me on as I experienced my first moments in space.

I was playing better and better, getting lots of attention in the papers and, eventually, from some NFL scouts. In my last season at Richmond, the team was 8–3 with a playoff bid. I went on to become an NCAA 1-AA Academic All-American and Richmond's career leader in receptions and receiving yards. I had caught a pass in each of the thirty-nine games in which I played.

As the 1985 football season came to a close, I got an unexpected call telling me I had been selected to apply for the Rhodes Scholarship. I would be the only student that the University of Richmond would submit that year. *Seriously? This can't be happening,* I thought to myself. Rhodes Scholars—thirty-two of them from the United States in any given year—get a free ride to Oxford University to attend grad school in the area of their choosing. Three years before, I was nearly expelled for violating the school's honor code. Now, I had been nominated for what was perhaps the world's most prestigious academic award. My success in the classroom and on the gridiron had drawn interest from outside the university. I had never heard of the Rhodes Scholars program, but the more I learned about it, the more I thought it would be something good for me to pursue. If the opportunity came, great. If not, there were other exciting possibilities, like professional football.

As luck would have it, the timing couldn't have been worse. The famously torturous Rhodes Scholarship selection process took place over the long Thanksgiving weekend, right as the Spiders were contending for a spot in the playoffs. NFL recruiters were calling to say they would be showing up in the next few weeks to check me out. And first semester finals were only a few weeks out.

Becoming a Rhodes Scholar requires surviving an application process that takes nearly as much resilience as vying for a spot in the Astronaut Corps—thankfully it's over more quickly. Rhodes Scholar candidates are nominated by their university, recommended by a minimum of five professors, advisors, or coaches, and asked to write a one-thousand-word essay with very high stakes. They then must get approval from their state committee, interview locally, shine at a cocktail party, interview on the state level, shine at yet another cocktail reception, and then interview at the regional level. If you survive all that, you move to the next level. I was one of 1,143 applicants from around the country that year, but I was eliminated somewhere between the first cocktail party and the second interview.

• • •

I remember walking by the student commons during lunchtime on my way to the chem lab. It was January of my senior year, and I was feeling the pressure of all the work I had to complete in my last semester. There was a television attached to the wall as you entered the commons from the dining hall and the news was carrying a story about the space shuttle *Challenger* launch. I was vaguely aware there was a launch scheduled that day, because the media had been chattering about NASA sending a teacher—Christa McAuliffe—into space for the first time.

And then it happened—tragedy. Seventy-three seconds into the flight, *Challenger* exploded 48,000 feet over the Florida coast in view of millions around the world. I saw the clip—over and over again, as the station replayed it constantly. Everyone was stunned. As I watched the tragedy unfold, I said to myself that, as a scientist, I would help prevent disasters like that from happening again. The vow didn't make much sense as my future still seemed earthbound.

I Get a Call

I was sitting in my room in Thomas Hall when the phone rang around eleven at night. It was April 30, a Wednesday, and I had just taken my last final at the University of Richmond. There were still a few weeks until graduation and people were pouring out of their rooms into the halls, looking for a spontaneous party or just to blow off steam. Earlier that day, Joe Bushofsky, a scout for the Detroit Lions, had phoned to ask me whether I'd be home that night, and I'd told him I would be. So did the scout for the Denver Broncos. Who goes out on a night the National Football League team might call?

The NFL Scouting Combine is the yearly scouting event held every February where players get a chance to show their skills to impress and raise their stock for draft day. I had not been invited, but Joe knew my stats, and knew that during my four years the team had gone from 0–10 to 8–3. Joe had flown down to Richmond a few weeks earlier and meticulously timed me in the 40-yard dash and ran me through a battery of agility drills. The Broncos had sent their scout too, as had the Dallas Cowboys. I hadn't heard back from the Cowboys, but the Broncos had indicated they were more interested in me as a free agent than a draft pick.

I picked up the receiver.

"Leland Melvin?" It was Joe. "You have been drafted in the eleventh round of the NFL college draft. Do you want to play for the Detroit Lions?"

I had made it to the NFL.

"Yes, sir," I said. And I hung up the phone.

I'm even-keeled. No super highs or lows. Instead of jumping up and down over this news, I immediately began to get ready for it. Even though I initially hadn't pictured myself playing professional football, I did see myself succeeding. My parents' culture of expectation, supplemented by my coaches' and professors' patient tutelage, continued to nourish and shape me as I prepared for my next challenge.

The next day, the news of my draft had spread across the University of Richmond campus. Local TV news carried the story, as did the *Richmond Times-Dispatch*. School administrators and professors whom I'd never met beamed with pride during news interviews. My teammates seemed ecstatic, and that was probably the best part of the whole thing.

My parents were elated for me, though they maintained the same composed demeanor they had throughout my childhood, a quality I had always admired. I'm sure my mother was warming up her prayer book in hopes of keeping me from getting even more badly injured in the NFL than I might have in NCAA football.

Four days later I was on a plane to Michigan for the Lions' minicamp. The week of intensive training gives the coaches a chance to assess the new players and decide who's coming in and who's not. It's nerve-racking, especially for a guy from the eleventh round, but things don't easily faze me. On the last day of practice, on the last play of the day, I caught a pass in the end zone—right in front of head coach Darryl Rogers. I finished on a high note, suddenly more confident that I would avoid being cut and find a place on the team. Brimming with self-assurance, I caught a plane back to Richmond to graduate from college.

Graduation was a blur. I remember feeling the peacefulness that comes from success, combined with excitement about the future and playing in the National Football League. During my four years at the university, I had reached the lowest and highest points more than once. I never once thought of giving up, even in the face of ridicule for our team's losing record and the suspicion of those Honor Council members who thought I had cheated on a test to get ahead. Instead, through the grace and wisdom of the good people in my life, I had developed a growth mind-set and truly believed anything was possible. Sitting in the Robins Center in my cap and gown, I never imagined that twenty-two years later I would stand on that very same stage and deliver the University of Richmond's 178th commencement address—three months after I'd returned from my first mission to space.

4

A Short Life in the NFL

The NFL Scouting Combine, the annual beauty pageant from which the NFL teams draft players, is an invitation-only event, held every February. It gives a once-in-a-lifetime opportunity to those who are on the list—and puts guys like me who weren't invited at an immediate disadvantage. Of the roughly 70,000 young men who play college football on any given year, only about 330 are chosen to attend.

The combine includes tests like the 40-yard dash and bench press, and gives coaches the opportunity to interview prospective players. But making it into this pool does not guarantee that a player will play on an NFL roster. Like NASA looking for new astronaut candidates, coaches are looking for specific skills and abilities to round out their teams, and hugely talented players sometimes don't make it if a team finds a player with a more compatible set of skills.

The event's organizers are often criticized for its reliance on the 40-yard dash and other speed drills, as well as bench press and other strength tests, because critics argue they weed out players with plenty of potential. The combine also measures arm and leg length and joint movement.

David Epstein, an investigative reporter and author of *The Sports Gene*, a book on how teams select elite athletes, questions the usefulness of the combine. He argues that the combine can end up overlooking certain qualities that make a player great. "One of the many failings of the NFL combine that tests prospective draft picks in physical measures is that arm length is not taken into account in the measure of strength," Epstein writes, for example. "Bench press is much easier for men with shorter arms, but longer arms are better for everything on the actual football field."

Nor, one might add, do those kinds of tests measure what some scouts call the intangibles. Those qualities, such as "high football IQ," field awareness, and the willingness to do the little things that help teams win, are often more detectable in game situations, not drills. The combine's shortcomings notwithstanding, the NFL always recommends that any player who dreams of playing professional football have a backup plan in case his hopes get dashed.

As I've said earlier, I didn't always dream about being in the NFL. Nor did I have that single-minded focus on playing football in college like most of the players I met. I had been a good football player at a young age, a bit on the small side but fast and able to catch pretty much anything. When I was a kid in the 1970s, soccer had not yet transformed the landscape of childhood sports, and wouldn't for another decade. Boys played football or basketball, or both. Because my father had played football in college and later in the Air Force, it was preordained that I would play football too. Yet I had also been good at tennis—and would have much preferred to spend a Saturday afternoon on the court than a Friday night in the stadium. If one of my grammar school teachers had asked me what I wanted to be when I grew up, I probably would have said a tennis star. But in Lynchburg, Virginia, in 1982, football meant scholarships. Astronauts were not yet part of my imagination.

• • •

I arrived in Rochester, Michigan, for the Lions' two-week training camp in July 1986. Being a rookie had its challenges off the field, but on the field we were all trying to impress, and during the first week, one player after another suffered injuries. League rules said you couldn't get cut if you were injured, and so players would often exaggerate injuries just to live another day. I remember sitting in the hot tub in the evenings listening to players share stories of the hardcore painkillers they depended on and it dawned on me that this was a different world than Richmond. The veteran players feared being replaced by a rookie, and the Lions coaches would ignore players' complaints of injuries and send them on the field just so they could cut them.

Back in the 1980s, no one was talking about concussions. If you got knocked out, you simply got back up. "This is a man's game," they would tell you, "you gotta man up." I remember once when we had a game at Virginia Military Institute (VMI) and it was raining and muddy. Greg Hasty, another tough

Georgia boy, had an abscess in his leg. It was infected. At halftime the doctor cut that thing out with a scalpel and bandaged him up and he went back out on the field with the dirt and mud. I remember looking at his bandaged leg with the blood streaming down. You were expected to be a gladiator—and that was just college.

In the NFL, coaches look for "the right stuff" in players, just like NASA does in astronauts. Performance on the field counts, along with a willingness to endure pain and injury to get the job done. Just like in the Astronaut Corps, the right stuff in pro football means never showing weakness. The idea is to make it look easy while displaying the level of skill that takes years to develop.

The science behind the ability of elite athletes to push through pain, or perhaps not even feel the pain, has been studied for decades, and the research falls on every end of the spectrum. Certainly many people have had the experience of being injured in a fall or an accident but not feeling the pain until after the emergency was over. In *The Sports Gene*, Epstein compares "stress-induced analgesia"—the brain's temporary blocking of pain in stressful situations—with what happens to athletes in competition, which is why it's imperative to have referees there to stop the play. The brain temporarily blocks the pain to accomplish what the person needs to get done.

But recent research takes this even further, suggesting that elite athletes simply feel pain less than nonathletes, allowing them to push their bodies to almost unfathomable extremes day after day in training situations. Without getting into a fruitless debate over what traits people are born with and what they learn, let's just say NFL players arrive on the field with a certain ability and willingness to endure pain—

including but not limited to exhaustion, trauma, and acute physical pain—that others just don't have. Consider also the endless combinations of family, social, or even financial pressures that contribute to a player's motivation to play, and you see how complicated the debate can get.

In my case, I had always had a mastery of mind over matter and an ability to endure a pretty high degree of discomfort every day to reach a goal way out on the horizon. I figured I was just born this way, or that it happened pretty quickly after I arrived.

But behind the stats and studies, there are stories. They tell us that success depends not just on individual skill but the strength and variety of the network supporting that individual. In my case, getting to the NFL resulted from the invaluable help of my parents, my neighbors, college teammates, professors, and coaches. They envisioned possibilities for me that I wasn't always aware of myself. Each, in some way, trusted in me enough to give me a chance.

During the second week of practice, I was running a route down the sideline accelerating to catch a pass when I felt it—a pull in my left hamstring. I winced and stumbled to the grass. In what would prove to be a stroke of luck, the trainer saw it and he pulled me out.

That same trainer vouched for me later when Coach Darryl Rogers and his staff discussed which players were injured and which were dogging it. Torn hamstrings can take as long as a year to heal, but I had a few weeks at the most to take it easy. By the end of July, I was training again, and playing well despite being in considerable pain at times. Coach Rogers came into practice and we had a team meeting. "A lot of you guys are in here slacking, trying to make the team by getting on injured reserve," he said. "If you don't get on the field, we're gonna cut you."

I was thinking, *I'd better get out there*. I was not fully healed enough to really be effective, but after hearing this charge from the coach, I felt I had to give it my all. Chuck Long, a rookie but a first-round draft choice, was at quarterback. I got in the game and ran a 15-yard hook. He threw me the ball and I caught it. I remember seeing the position coach and he was really excited. I caught two passes in the game against Philadelphia and another two against Seattle. There was another play when we were at our 40-yard line, man-to-man coverage. I was running down the sideline trying to lose my defender, but I just didn't have the juice. I dove but the ball was way in front of me. The coach said, "Don't ever do that again. You're gonna hurt yourself." It felt like I was moving in slow motion a little bit and I knew it: My leg was simply worn out.

A few days after the fourth preseason game, on a Tuesday, someone knocked on my door. I recall it was about 6:45 a.m., and I knew what it meant. Everybody knew that this week would be the final round of cuts before the season started. For reasons beyond explanation, in the NFL the coaches always cut players on Tuesdays.

My roommate at the time was Lyle Pickens, a rookie defensive back from the University of Colorado. Lyle had been picked up in the ninth round of the draft, and reaching the NFL had been his life's goal. He heard the knock and, like me, he knew what was coming. "Hey, Coach wants to see you guys. Says to bring your playbook," the guy barked through the door. I was on a plane home to Lynchburg that afternoon.

My best friend on the team at the time was another rookie wide receiver, Allyn Griffin, drafted from the University of

Wyoming in the eighth round of the 1986 draft. Allyn was released from the team around the same time as Lyle and me, and he recently told me about the conversation he had that summer with Coach Rogers. Allyn was summoned to meet with him in the small room at Rochester College that was the coach's makeshift office during training camp. As the two men sat across the table, Allyn tried to stay composed. For about twenty minutes they talked about the team, the players, and how Allyn felt about the training, and then Rogers finally got around to the question he'd brought Allyn there for.

"Do you think you belong in the NFL?" Rogers asked him. To which Allyn replied, "I don't know." When Allyn recounted this story nearly thirty years later, he seemed haunted by regret. "Why didn't I just say yes?"

It was different for me. I got word of my release from Joe Bushofsky, a man with a big heart. I could tell he felt almost as bad as some of the players he released. I guess I was lucky—he didn't ask me to weigh in on it. "You always try to be as kind as you can," Joe said, years later, long after he retired from scouting. "I would explain that the coaches feel the chances of them working with the team are not good. I would tell them right away, 'You're being released today.'"

Most of the players took the news quietly, Joe told me. "They would nod or drop their heads, trying to grasp that their future was not going the way they had dreamed," he said. "They would sign the release form and then leave, sometimes turning to thank me for the opportunity. For most of them it was not really a surprise, but that didn't take the sting away. Others sat there in stunned disbelief and argued that they hadn't been given a fair chance." Joe recalled one player who thought for sure he'd made the team. "When I told him he was cut, he got very upset," Joe recalled. "He began to cry. Then he pulled out a Bible and began to read from it."

For me, it was a blow, but I also knew that as an eleventh-round pick, and considering my injury, getting cut had always been a very real possibility.

Sometimes I find myself wondering about certain people in my life, like Allyn, who experienced disappointments and failures. He had a difficult time after his release from the Lions. He went home to Wyoming, where he sank into a depression and wandered for a time into substance abuse until finally turning it around. Today, he coaches high school football and mentors kids who face the same steep odds he did as a teenager. Maybe he would not be doing this important work if he had secured a career with the NFL. I like to think helping others realize their dreams is exactly where he's supposed to be.

There are times in our lives when we are told we're not good enough. Consider that some astronauts applied to the Astronaut Corps more than a dozen times before finally making it into the program. What if they had given up after the first rejection? Growing up I had never seen myself as a professional football player, though once it was within my reach, I moved heaven and earth to achieve it. Now I was returning to my hometown with no immediate plan. What now? I had no idea.

It turns out I wasn't done with the NFL. The following morning, I was lying in bed in my childhood room thinking about my next move, when I got a call from my agent, Will Rackley. Will wasn't really an agent. At least not yet. He just wanted to be, and I was his first client. He owned a business in Richmond, but football was his passion. He was a talker. So I figured he could talk up a deal for me.

"The Dallas Cowboys want to check out your leg," he said. "They want you in Dallas tomorrow." I got on a flight that morning and by the afternoon I was catching passes from Pro Bowler Danny White. Still, I flew home on Thursday and packed my bags when Rackley said the Toronto Argonauts were interested in having me train with them for two weeks so they could evaluate my skills and assess my rehab. *Canada?* I didn't know much about the Canadian Football League, but I had a friend on the team from the University of Richmond, Mark Seal. Without any word from Dallas, I had no other options, so I caught a flight to Toronto that Friday morning. Detroit, Dallas, and now Toronto—all in less than a week.

A few days into the Argonauts' practice, I got another call from Rackley. "Dallas wants to sign you," he said, excitedly. The Cowboys wanted me as a free agent for the 1987 season. I was back in the NFL. Once again, I had been given a second chance.

• • •

It was only September and the Cowboys' minicamp didn't start until March so I took a job working for Rackley, delivering packages from his office in Richmond. One afternoon I bumped into Dr. Raymond Dominey, the husband of one of my chemistry professors at the University of Virginia. "Why don't you go talk to Glenn Stoner, at the UVA materials science engineering department?" he said.

"Why would I do that?" I asked. I was biding my time before playing for the Cowboys.

"Just go and see what he has to say," he replied. I drove to Charlottesville the next day. Years later, Dr. Dominey told me he had a hunch materials science would resonate with me because it combined chemistry and engineering, and he was right. I would soon learn that Dr. Stoner was both a brilliant scientist and a football fan, so I think the idea of an NFL player on his staff was too much to resist. On that day, Dr. Stoner offered me a job as a research assistant until the start of mini-camp, which sounded a lot better to me than delivering packages. It paid better too. I rented a room in an apartment on Fifteenth Street in Charlottesville and during my free time worked out with the school's football team.

I felt right at home. The materials science engineering department might sound like a nerd convention, but the students there were determined to have fun. They all seemed to love the work they did.

Before long, Dr. Stoner encouraged me to apply to the graduate program even though I was leaving for the Cowboys' minicamp in a few months. I was reluctant at first, knowing what that would require—long hours of studying on top of playing professional football. But I also had just learned the painful lesson that a coveted NFL career can end before it even starts. Having a backup plan couldn't hurt. I was a free agent, which meant there was no commitment by the team, nothing close to a guarantee I would win a place on the Cowboys' roster that fall. The NFL was a gamble. Getting a graduate degree allowed me to hedge my bets.

I was accepted for the spring semester and started classes in mid-January just before the Cowboys' camp was scheduled to start. Luckily for me, the engineering program had some classes that were broadcast for remote students. I was not able to attend the classes at the designated times, so the department videotaped my math for materials science and crystallography courses. I could call Dr. Stoner or any of my colleagues in the department whenever I needed help.

Balancing the NFL and grad school proved to be the hardest thing I'd ever done. We spent ten or more hours a day conditioning in the gym and running drills on the field. Four of those hours we spent lifting weights. One day I was lifting with a lineman, a first- or second-year guy. As he was spotting me on the bench, he turned to Tony Dorsett, who was walking by. He said, "Tony, we're going to have a good season." Dorsett, a legendary running back, nodded without missing a beat. He was soon traded to Denver and lasted another year before torn knee ligaments ended his career. Herschel Walker was also there, and I remember him being one of the most positive and encouraging people on the team, always calling the new guys rookies and then laughing with his big southern Georgia smile.

We would scrimmage in helmets, shoulder pads, and shorts. Technically, there was no tackling. Once I ran a 15-yard crossing pattern and collided with Bill Bates, one of the hardest hitters in the league. He laid the hammer down on me. He was very apologetic, picking me up and telling me he was sorry. Players who had gone to big colleges, like the University of Southern California or the University of Michigan, could come right into the league and make an impact right away. In that sense, the college powerhouses are like the NFL. Having competed in college against Virginia Tech and other big-time programs, I had already been introduced to the NFL's speed of play.

When we weren't training, we ate. We ate a lot. But while most of the guys would do little else besides working out and consuming food, I would also do hours of schoolwork every night. I would get back to the apartment I shared with Fran McDermott, to watch videos of the materials science classes I had missed, and then catch a few hours of sleep before getting up the next day to head straight to the gym. I took my exams proctored by a professor at nearby Southern Methodist University.

By the time I headed home for a short visit before the team moved training to southern California, I had completely transformed my body. I had bulked up so dramatically that when I pulled up to the house and got out of my car, my parents hardly recognized me. My mother was visibly upset that I had changed so profoundly in only four months. At the University of Richmond I had been in top physical condition, but in the pros I had reached a new level, that of a perfectly tuned machine.

Back at the apartment, I was faced with a dilemma. How could I learn the Cowboys' plays when I needed to spend every spare minute studying? The Cowboys' playbook was a sophisticated three-inch-thick document, with every page containing diagrams and detailed descriptions of plays. I had learned some programming at a minority introduction to engineering program I participated in during high school. So, relying on my knowledge of Fortran 77 and equipped with a new Zenith laptop computer, I programmed electronic flashcards of the major plays. It was a crude system—nothing I would ever be able to sell to Microsoft—but it got the job done. One of the linebackers heard about my flashcards and asked, "Hey, man, can you do that for me? I'm having trouble."

I wanted to help him, but with the demands of school and an NFL training camp, I didn't have the time to do all the work needed to develop a new set of flashcards. I told him I could give him the code so that he could do his own programming, but nothing came of it.

• • •

I had been stretching on the Cowboys' practice field when Danny White called over to me. "Hey, Rookie, let's throw some," he said. It was April, nearly five months before the first game. We were in Valley Ranch at the Cowboys' training facility. Danny knew that a chronic hamstring injury had sidelined my career with Detroit, and we agreed I would run at half speed during the exercises to help me get loose. The play we decided to run was a 15-yard out downfield before angling back toward the sideline.

All that changed when Tom Landry walked onto the field. Anytime a quarterback has an opportunity to impress the head coach, he's going to take it. Danny changed the play, did an audible, and sent me deep. I knew Landry was watching, and I thought, *If he sees me running half-speed and not go after the ball, he's going to see me as an unmotivated rookie.* You never get a second chance to make a first impression. So, I kicked it into gear, giving it everything I had. I never made it to the end zone. I felt this stinging pain in my upper left hamstring. I came to a stop, holding my wounded muscle. Danny ran over and told me he was sorry. Landry didn't say a word.

My decision to go for it put me back in the same spot as in Detroit—another injured rookie trying to make the team. I went to the trainer. The first thing they do is put you on the table, look at the muscle, and put ice on it to minimize swelling and inflammation. Typical treatment for hamstring injuries involves rest, ice, compression, and elevation. The

Cowboys' trainers were always looking for new ways to get guys back on the field, so they sent me to an acupuncturist in Dallas. I also worked with a physical therapist and a trainer as intensely as I'd been working on the field. I was determined. The acupuncturist inserted a needle directly into my nerve bundle, relieving tension and stimulating my muscle until it twitched like a frog leg in a biology experiment. I had ten sessions over a two- to three-week period. The program worked brilliantly, enabling me to fully recover. Soon I would be back to the team's hardcore practice schedule, in plenty of time for the start of preseason.

During my recuperation, I visited my professors at the University of Virginia and my parents in Lynchburg before heading to Cowboys' training camp. At camp, I looked forward to getting started when Coach Landry summoned me to a conference room. "And bring your playbook," I was told. I took that as a good sign this time. The only time you're asked to bring your playbook is when you're getting cut or switching positions, and I knew Landry wouldn't cut me before evaluating my leg. He hadn't seen me on the field much since I had recovered and gained my strength back. I was in top shape. I guessed he had decided to move me from wide receiver to defensive back, which did not seem unreasonable.

With rookies Landry was very distant. Maybe it was different for first-round draft choices and other top players, but when you're a free agent you're like a walk-on. I think he may have nodded his head and grunted at me one time. The real communication and interaction typically came through position coaches. I walked into a classroom carrying my playbook and saw Landry sitting at a long table. For once, he was looking squarely at me. "Leland, I'm releasing you from the team," he said. And with those few words, my NFL career was over.

I can't say I wasn't angry. I felt betrayed. I hadn't been given a legitimate chance. That afternoon, I packed my things and moved to my uncle's apartment in Los Angeles, where I spent a few days sitting on the beach, contemplating what had just happened. My first thought was to keep fighting, to find myself a spot somewhere else in the NFL. I had come this far. But my instincts told me otherwise. And so just as I had done at every other turning point in my life, I asked God for direction.

Soon it became clear that it was time to turn the page and start a new chapter in my life. I believe that everything happens for a reason, and it was slowly becoming clear to me what I was supposed to do next. I took a flight to Lynchburg and moved into an apartment on Jefferson Park Avenue in Charlottesville for the fall semester in the university's graduate school of materials science and engineering. I got there just in time for the start of classes.

Back to Grad School

During the off-season, a number of pro football players attended law school or business school. A lot of them were really sharp guys, but none of them were enrolled in the University of Virginia's graduate school of materials science and engineering. For the first few days of the semester I was surrounded by guys wondering what it was like to play for "America's Team." Only a week before, I had been training with the Cowboys, and my classmates found that endlessly amusing. And then there was Marlene.

"You played pro football?" she asked in a thick Boston accent. I was pretty certain she knew nothing about football. "What team?"

I told her.

"What position?"

"Wide receiver."

"I guess that means you're really . . . fast?" she asked, raising her eyebrows. I wondered if she had a clue what a wide receiver did.

"Uh, yeah . . . I'm pretty fast, I guess."

That exchange with Marlene got me thinking. It was a relief to meet someone from a different world than the one I had been living and breathing for more than a year. And with that, I closed the chapter on my life as a professional football player.

• • •

At the university, Marlene and I formed a study group with Brenda Jones, the student who had been sending me her class notes when I was with the Cowboys. We called ourselves "The Three Musketeers." Without them I don't think I would have had nearly as much fun as I did in Charlottesville. We often gathered for softball games and happy hour at local pubs. I found a family in the Materials Science Engineering Department, and later on would invite Dr. Dominey and Dr. George Cahen, another of my professors, to watch my first shuttle launch at Cape Canaveral. Marlene and Brenda also showed up, though I didn't see any of them because shuttle astronauts are in quarantine by the time their guests arrive to see the launch.

Most of my colleagues went on to careers in manufacturing or product design, but Marlene followed a different track. She got a degree in electrochemical corrosion and ended up teaching science in the Boston public school system. Many of her middle school students came from public housing in dilapidated neighborhoods and faced steep odds in their paths to success. Years later I spoke to them about the importance of staying in school. I wanted them to know that failures in life are the building blocks for later success, and that anything was possible. Even my serendipitous path to space.

5

NASA Langley

A few times a year, the Graduate School of Engineering would host a career fair near campus at the Omni Hotel. It was early in 1989, and I was planning to get a job at one of the top chemical companies in the spring when I finished my master's degree program in materials science engineering. I had met with representatives from DuPont and Dow Chemical Company, and both were interested in interviewing me at their home offices. The fair was coming to an end and as I made my way to the exit, I saw a logo above the booth in front of me. It read "NASA." I kept walking. It was getting late and the bear in my belly was starting to growl. I was thinking of calling Marlene and BJ about Thai food when somebody called out to me.

I turned to see a diminutive black woman, all of five foot

two with a big smile and bright eyes. Her badge said Rosa
Webster, NASA Langley Research Center. She grabbed my
arm. "What's your name?" she asked. Despite her small stat-
ure, she had a huge, larger-than-life presence and a pretty
strong grip. I told her my name and she said, "Leland, you're
going to work at NASA. I've been looking for you all day."

Rosa then asked me to help her move some of her boxes to
her car, and I stuck around after she started breaking down
her booth. All the while, I was thinking, *Who is this woman in
my face?* And secondly, *I'm not working for NASA.*

Who Works at NASA?

By the late 1980s, NASA had made considerable strides in
hiring black scientists and engineers, but Langley director
Paul Holloway was under orders to further improve diversity
at the space agency. Rosa was one of a cadre of employees
whom Holloway had assigned to find and recruit minority
candidates. Someone in the university's engineering department had supplied Rosa's boss, Charles Blankenship, with a
list of graduate students and my name had been on it. "We
were looking for really strong candidates with strong creden-
tials who we'd have a hard time getting our hands on because
of our disadvantages to private industry," Rosa later told me.
"I had spent the whole day thinking, *Is Leland Melvin going to
show up?*"

Working at NASA had never crossed my mind. I mean,
who works at NASA? Certainly, nobody who looked like me.
Instead, like most of my grad school buddies, I had set my
sights on private industry, where I knew the pay would be
better and I would have more opportunity.

About a half hour after I first met Rosa, she said NASA would send me a written job offer within a few days. *A job offer?* I was puzzled. During that time, NASA leadership gave certain staffers, including Rosa, the authority to make verbal job offers on the spot, particularly at job fairs where there was fierce competition for qualified minority graduates of science, technology, engineering, and math programs. And they were probably right to do so. In the days following the job fair, I visited Dow Chemical and DuPont headquarters, where I got a taste of the kind of jobs I always expected to be offered after graduate school.

Soon after I got back, NASA Langley's offer letter was waiting. I thought of Rosa as I picked up the letter. She had pitched NASA hard as I carried her materials from the job fair to her car. If her gig as a physicist somehow went south, she'd have amazing opportunities in sales. The letter included an invitation to come to the agency. The very least I could do was to go visit.

• • •

NASA has nine research centers distributed all around the country. Centers like NASA Langley do basic and applied research, in contrast to operational centers like Johnson, which focuses on astronaut training, and the Kennedy Space Center, where the mission is launching people into space. NASA Langley is often confused with CIA headquarters in Langley, Virginia, but it's actually in Hampton, Virginia, and shares a runway with the Langley Air Force Base. My new job was about three hours from my hometown.

On my first day, we went into this room where they gave us visitor badges and told us about the center and explained our benefits package. There were three newcomers, one of whom was a friend of mine. Four black graduates from UVA eventually took jobs at Langley, although we didn't all come in at that same time. We all knew each other, though, and we were probably all recruited by Rosa Webster. All of us eventually reached leadership roles.

After orientation, I went to my office in the Nondestructive Evaluation Sciences Branch located in Building 1230, one of the oldest buildings on the campus. I talked to the branch chief and the man who would become my first boss, Joe Heyman. Heyman was a physicist who had a reputation around the space agency for getting his projects funded. A former used-car salesman with a PhD, Heyman was also very welcoming and encouraged his staff to work in a collegial fashion. Having lived in the South, I was wary of being the first African American in the branch. Heyman was from New England, and he had a way of embracing everyone. Plus his team included other progressive thinkers. I liked that Heyman believed scientists did their best work when unencumbered by bureaucracy. He also spoke eloquently about NASA being a family and a family-friendly work environment. Heyman reminded me of my father.

The visit proved to be a major turning point in my life. Rosa had been right—NASA Langley was a good fit for the kind of work I wanted to do, and my goal of the big salaries, benefits, and bonuses that came with working in corporate America fast became a distant memory. I was headed to NASA.

At the space agency, I spent my time doing research on optical systems that use lasers and optical fibers to remotely detect damage in aerospace structures and vehicles. NASA used these sensors to help improve the safety and reliability of vehicles, like the space shuttle. The shuttle's fragile protective tiles can withstand 3,000 degrees Fahrenheit, and inspection systems that quickly assess their integrity can save time, money, and lives.

Our work, unfortunately, could not prevent the 2003 tragedy of the *Columbia* mission when a brick-sized piece of foam dislodged from the shuttle's fuel tank during takeoff and damaged the tiles on the left wing. Once in orbit, there was no way for the crew to inspect the damage on the outside of the spacecraft except for what they could see using cameras and binoculars. After *Columbia*, my work in sensors helped NASA shape new methods of scanning tiles and wings while the shuttle was in orbit to help prevent future disasters.

During my first three years at Langley, I did fiber optics research with Jim Sirkis, a professor at the University of Maryland. He came to Langley on a sabbatical, conducting research on fiber optic sensors. We started working together to make these sensors. Jim had a PhD in mechanical engineering and I was just breaking into this area of research, but he was personable and we got along well.

While recruiting me, Rosa had told me about the Graduate Education Program (GEP) for NASA employees. She told me, "You can get your PhD, we'll pay for it and you'll be on salary." That only aroused my curiosity about becoming "Dr. Leland Melvin."

When Jim encouraged me to come to the University of Maryland to pursue a PhD, looming budget cuts and the possible elimination of the GEP spurred me into action. But, first I had to take some prerequisite courses. For a year I took undergraduate classes in statics, dynamics, strength of materials, and differential equations across the bridge from Langley at Old Dominion University in Norfolk. NASA was great about allowing you to study and work at the same time. I would leave during the workday to take classes and go home to study.

After completing my courses, I packed up my stuff and moved to College Park, Maryland. I was now taking not only mechanical engineering classes but also electrical engineering. With a year of courses under my belt, I began to realize my heart really was not in the PhD program. I had forced myself into it because it was something I told myself I wanted to do. People told me, "You should just stick it out. You're being paid your full salary and your tuition is getting paid. Why would you turn that down?"

It was true. I had an ideal situation with friends and people who cared about me. Still, something didn't feel right, and I decided to trust my instincts. I returned to Langley to a job that would provide access to the space program and put me on my path to the stars. But I didn't know that then. All I knew was that I needed to be still and quiet and listen to the universe. It was time to move on and return to NASA.

I became the program manager of the optics lab at Langley. We developed glass fibers that functioned much as nerves do in the human body, detecting leaks and other forms of damage in the reusable launch vehicle designed to replace the aging space shuttle. Our work was part of an effort to update vehicle technology while also reducing operational costs and turnaround time. Each fiber is about 125 microns in diameter, roughly similar to that of a human hair. They could do the inspection work of an army of techs, helping astronauts get to space and back faster and more efficiently.

Race, Langley, and Katherine Johnson

Shortly after returning to Langley, I joined the National Technical Association (NTA), the oldest group of black scientists and engineers in the country. It was at an NTA gathering where I first met Katherine Johnson, the famed African American mathematician, physicist, and space scientist. She had officially retired from NASA in 1986, three years before I arrived. We exchanged pleasantries, and I recall that for her age she had such a presence. At the time, I had no clue about the woman I had just met. It would take time for me to learn about her true significance to America's space program.

Katherine Johnson is a living legend at NASA and, unfortunately, someone who had escaped public attention until very late in her life. Her name doesn't grace many history books, her story is still missing from many school lesson plans, and her accomplishments aren't praised enough during Black History Month—or any other time of year for that matter.

We as a nation are the worse off for it.

I wasn't alone in not knowing Katherine's contributions to science and society. Years after that technical association meeting, I had the opportunity to introduce Katherine to famed singer, songwriter, and producer Pharrell Williams at an event in Virginia Beach for the From One Hand To AnOTHER program. The program is a local nonprofit foundation Pharrell set up to provide science, technology, engineering, arts, math, and motivational-related tools to youngsters to ensure their success as adults. The two shared a few words, but the introduction was forgotten. Pharrell would recall it when he began work as a producer on the film *Hidden Figures*, the bio-pic depicting Katherine, Mary Jackson, and Dorothy Vaughan—three brilliant black women who made history at NASA. I had to laugh at the movie premiere when Pharrell reminded me that he had met Katherine before he knew her story.

"I remember seeing her," he said. "But I didn't know she was all that."

Katherine was all that and then some. She was the one who calculated the first orbital flight for the late John Glenn, the American astronaut credited with restoring the nation's confidence in space travel during the early space race with the Soviet Union. Glenn believed in Katherine's abilities as a mathematician so much that he requested her to check the work of the IBM computers NASA had used to calculate Glenn's spaceflight.

For Katherine, the trailblazing career at the space agency started with counting. As a child, Katherine counted everything in her path, from the steps leading up to the church, to the dishes she washed, to the stars she saw in the sky. Her parents knew she was different. Unfortunately, for African Americans in Katherine's hometown of White Sulfur Springs, West Virginia, school only went through eighth grade. Determined to make sure Katherine received a decent education, her parents agreed that she and her mother would move 120 miles away to the nearest town where Katherine could attend high school. They made that trek across a landscape littered with the Ku Klux Klan and onerous Jim Crow laws.

Katherine sailed through high school in Institute, West Virginia, graduating at fourteen, and then went on to get a degree in math from West Virginia State at only eighteen. There was no holding her back—and thankfully nobody at NASA dared try for long.

Katherine started out as a teacher, but she heard that a nearby aeronautics laboratory was looking to hire black women mathematicians to do calculations. In 1953 she became "a computer with a skirt," one in a legion of women hired to perform calculations—before the days when electronics were ubiquitous.

Katherine's genius for calculations defied the norm, even at NASA. She calculated the launch window for Alan Shepard's Mercury mission that marked America's first manned trip to space, and assisted in calculating John Glenn's orbit around the Earth. In her later NASA career, she worked on the space shuttle program and the Earth resources satellite. After retiring, she encouraged students to pursue careers in science and technology fields. Katherine's contributions to NASA, however, went far beyond science. Perhaps her greatest gift involved her willingness to ignore anyone who attempted to stand in her way because of her race or gender. Charles Bolden, the first astronaut to head NASA and the agency's first African American leader, took note of her determination. She had once said, "I'm as good as anybody, but no better," he recalled. "The truth, in fact, is that Katherine is indeed better. She's one of the greatest minds ever to grace our agency or our country, and because of the trail she blazed, young Americans like my granddaughters can pursue their own dreams without a feeling of inferiority."

Langley itself is like a small town. After work my colleagues and I would stay together through the evening, playing basketball or softball together or meeting to discuss our work or hear presentations. Though she was retired, Katherine would show up and plant herself at the center of the discussions and events. It was like a family and Katherine was the matriarch. She was brilliant and inspiring, and seeing her command of her environment and the precision she brought to every endeavor served as an example at a time when I was still trying to find my way.

In 2015 I had the pleasure of accompanying Katherine to the White House to see President Barack Obama award her the Presidential Medal of Freedom. "In her thirty-three years at NASA," he said, "Katherine was a pioneer who broke the barriers of race and gender, showing generations of young people that everyone can excel in math and science, and reach for the stars." During the ceremony, I thought back on how much my success depended on the vision and perseverance of people like Katherine. Individuals with grit and determination—like Dr. Walter Johnson, who helped Althea Gibson and Arthur Ashe break the color barrier in tennis, from my old neighborhood on Pierce Street—came to mind.

More recently, I had another opportunity to be with Katherine in Hampton, Virginia, at a sneak preview of the film *Hidden Figures*, starring Octavia Spencer, Janelle Monáe, and Taraji P. Henson as the young Katherine. It was quite a tribute to the ninety-seven-year-old Katherine and her two African American colleagues, Dorothy Vaughan, NASA's first black manager, and Mary Jackson, the space agency's first black female engineer. The three were the brains behind some of NASA's most important early missions, and it was so humbling to see Katherine relive her time at the space agency. She was smiling during the whole movie.

Katherine didn't march on Washington, but her brilliance

at NASA, the sheer strength of her intellect, and the respect she demanded from her peers made her a revolutionary in her own right. She had far broader impact for blacks at NASA than I'm sure she knew at the time. In her early days at the space agency, I do not believe she saw herself as part of a movement, but as an individual demanding to be heard in a predominantly white-male organization.

The reality is that NASA played a major role in the government's efforts to integrate the South. In the 1950s and 1960s, the space race was accelerating during the Cold War just as the struggle for civil rights was erupting in the South. President Kennedy saw an opportunity in the expansion of the space program to try to break Jim Crow's tenacious grip.

In *We Could Not Fail*, Richard Paul and Steven Moss presented a study of the first African Americans in the space program and the national consequences of their inclusion. The authors argue that NASA's role in southern desegregation hasn't been fully appreciated. "The work NASA did as part of federal civil rights efforts, as well as the social consequences of its presence in the South," they contend, "bridges two great American stories of the early 1960s."

Another important character in America's space program emerged in 1966, when *Star Trek* creator Gene Roddenberry decided to include Nichelle Nichols, an African American actress, in the show's original cast. As Lieutenant Nyota Uhura, communications officer of the starship *Enterprise*, Nichols enabled television viewers to see a black woman in a television series who was not a servant. She had a role of authority.

"When I was nine years old *Star Trek* came on," the Oscar-winning actress Whoopi Goldberg once told a reporter. "I looked at it and I went screaming through the house, 'Come here, Mom, everybody, come quick, come quick, there's a black lady on television and she ain't no maid!' I knew right then and there I could be anything I wanted to be."

Nichols didn't completely grasp her character's significance until the conclusion of the first season, when she told Roddenberry she was leaving the show to take a part in a Broadway-bound production. Roddenberry was determined to keep her. He told her to take the weekend to think it over. That weekend, during an NAACP fundraiser at a Beverly Hills hotel, Nichols was told that a fan wanted to meet her. *Oh no, another Trekkie*, she thought.

Nichols turned to meet the fan and found herself looking into the smiling face of the iconic civil rights leader Dr. Martin Luther King Jr. "I was breathless," she said.

"Yes, Miss Nichols, I am that Trekkie," King told her. "I'm a *Star Trek* fan." *Star Trek*, King said, was one of the only shows that he and Coretta would let their small children stay up late to watch, largely due to the fact that the role of the ship's communications officer was played by a black woman. Nichols thanked King for the compliment. She also told him she was leaving the show.

Nichols later, during an interview with the *Washington Post*, recalled King's reaction. "He said something along the lines of, 'Nichelle, whether you like it or not, you have become a symbol. If you leave, they can replace you with a blond-haired white girl, and it will be like you were never there. What you've accomplished, for all of us, will only be real if you stay.' That got me thinking about how it would look for fans of color around the country if they saw me leave. I saw that this was bigger than just me."

Star Trek was canceled in 1969 after only three seasons,

but Lieutenant Uhura's capacity to inspire continued. NASA hired Nichols to help the agency recruit women and minorities into the space program, and that she did. Among the astronauts she helped recruit were Guion Bluford, the first African American in space, and the first woman, Sally Ride. While *Star Trek* motivated many people to come to work at NASA, few of its characters had such a direct and lasting influence as Lieutenant Uhura.

• • •

A few years back I had the honor of speaking on a panel at the National Air and Space Museum with Julius Montgomery, the first black engineer to work in the space program. Morgan Watson was also on the program. He grew up picking cotton before setting his sights on a career in science. He joined NASA in Huntsville, Alabama, and eventually worked on the launchpad for the *Saturn V* rocket that flew to the moon. Our third panelist was Dr. Mae Jemison, the first black woman to go to space. She arrived at NASA nearly thirty years ago after Watson joined the space agency.

Montgomery was an electronics technician, hired to work at Cape Canaveral on a NASA project overseen by RCA. It was 1956, and in those days there may have been no worse place to be an African American than Florida. Authors Paul and Moss wrote that a black man had a better chance of being lynched in the area around Cape Canaveral than almost anywhere else in the United States.

Racial segregation didn't dampen Montgomery's drive. Besides becoming the first African American to be hired as anything other than a janitor, Montgomery also helped found the Florida Institute of Technology, a one-time night school for

NASA employees that is now a respected university. To this day, the school gives out a prestigious annual award named for Montgomery.

When I met Watson and Montgomery in 2010, Montgomery was nearing ninety. Still, he remembered the terror he felt nearly fifty years ago when he first walked into that NASA lab full of angry white men who didn't think he belonged there. After the panel ended, I had a chance to speak with him during the reception. He came up to me and shook my hand. "You know," he said. "You astronauts, you're the bravest people I ever met."

I couldn't believe he was saying this to me, this from a man who opened doors at the space agency so I could someday fly in space. I'm sure the notion of a black astronaut at NASA was unthinkable during Montgomery's time at the agency. "No sir," I said. "I heard your story. You are the bravest person I ever met."

Reluctant Astronaut

Langley seemed far away from the action associated with being an astronaut. It had its own culture, its own politics. But in 1995 my research colleague Charles Camarda applied for the Astronaut Corps and, to my surprise, was accepted— and on his second try. He had applied unsuccessfully nearly twenty years earlier, when he was fresh out of college with a BS in aerospace engineering from Brooklyn Polytechnic Institute and only a few years' work experience. In the intervening years, he'd gotten a master's degree in engineering from George Washington University and a PhD from Virginia Tech. He also raised a daughter on his own. He'd waited until his daughter was old enough to understand the importance of space exploration before trying again.

Charlie is hardly the picture of a NASA astronaut. Standing at about five foot four, he sports a thick brown mustache and a thick Brooklyn accent. While most astronauts will tell you they had wanted to be an astronaut since they were a kid, Charlie would have told you he wanted to be a boxer. Second to that he'd probably have told you he wanted to be a research scientist, but the lure of the ultimate challenge proved too much to resist, and Charlie took the plunge and applied.

Charlie, an astronaut? If Charlie can do it, I can do it, I thought. But even at the urging of my good friend and Langley engineer Thomas Kashangaki, who actually gave me an application, I still didn't apply.

I'm not sure why, other than the odds seemed long and the process exhausting. Yet about a year later, Charlie flew from Houston to Langley in the sleek blue-and-white NASA T-38 training aircraft piloted by astronaut John Young, possibly the most decorated and accomplished astronaut in the space program. Young was a legend. He had flown almost every aircraft known to man and flew on every space vehicle from *Mercury* to *Apollo*. In April 1972, he walked on the moon. If that wasn't enough, he went on to command the first space shuttle mission STS-1, *Columbia*. He had the longest career of any astronaut ever, and yet here he was, visiting me at Langley.

The T-38 Young flew that day is an amazing aircraft, capable of flying supersonic up to Mach 1.3 and above 50,000 feet, 10,000 feet higher than the cruising altitude of a commercial jet. I would eventually come to know the T-38's mind-numbing acceleration at seven G's, seven times the force of gravity. On that day, though, I was drooling at the idea of riding in one.

During their visit, I had a chance to present to Charlie and Young the research on optical fibers I was doing to improve safety in the space program. Though I didn't know it at the time, Charlie had talked me up to Young the entire flight to Langley, knowing that Young's input in the selection process could prove invaluable. I was in awe of Young, but wouldn't you know it, he slept through my entire talk. Still, he was John Young so nobody dared rouse him. But when he finally woke up he turned to me and said, "Great job, Leland. You should apply to be an astronaut."

Years later, Charlie would go on to fly on space shuttle *Discovery*'s first mission after the *Columbia* disaster—known as the Return to Flight mission—a flight that took extraordinary courage and nerve, and Charlie would prove himself time and again to be an excellent mission specialist and engineer.

The requirements for becoming an astronaut have gone through several transformations since the first class was selected in 1959. Back then President Eisenhower required that all astronauts be military-trained pilots with a minimum of one thousand hours piloting jets. That rule was in place until the fourth astronaut class in 1965, when the pool included candidates chosen for their science and academic backgrounds.

I had finally applied, just before Young's visit to Langley. When I was selected, I learned that NASA, much like the National Football League, was looking for players with certain skills to give depth to the team.

Patience is the first requirement to become an astronaut. When the pool of 2,500 applicants had finally been whittled down to 120, I was beginning to think that I actually had a chance of making it in. The medical exam was straightforward and certainly more extensive than anything I'd encountered. There's a vision exam and a dental exam. An MRI looks for any undiagnosed conditions, and they check your heart and your cardiovascular health. There's a VO2 max stress test on the treadmill to check your endurance level. Every interaction with NASA personnel is an opportunity for the selection committee to better understand your character, through your treatment of others and your ability to get along. Being condescending to the janitor cleaning the bathroom would be a strike against you. I recall getting fitted for a flight suit and seeing Charlie Precourt, then head of the Astronaut Corps, listening to my conversation with the technicians.

Then the five-day interview process begins, culminating in an hour-long discussion with the selection committee, made up of senior leaders, administrators, and five fighter jocks, highly decorated Navy pilots who actually appeared in Tom Wolfe's book *The Right Stuff*. I don't usually get nervous but I was then. When the questioning started, a committee member had to catch me when I leaned too far back in my chair and almost fell over. My demonstration of "the right stuff" was off to a bad start.

Like the other candidates, I also had been required to write an essay explaining what I could contribute to the human exploration of space. I began mine by discussing the values I learned from my parents.

"At the age of five, my father drove me to my first Little League basketball practice," I wrote, "where he emphatically stressed for me to work hard, have fun, and share the ball. Those few simple words have resonated in my head countless times throughout my school years and professional life. Though simple, they emanated the commitment and selflessness required to work and play as a team. My parents taught me early on that virtues such as courage, integrity, and faith were assets that would guarantee success. By fully embracing these values they assured me that being myself would be enough. Therefore, I offer myself to be used further in the human exploration of space like so many others before me."

I didn't have to go quite as far back during the interview. The committee asked me to trace the path of my candidacy, starting with high school. I'm sure a lot of applicants talk about how they'd wanted to be an astronaut their whole lives, but that wasn't the case with me. Growing up, the space race was in full throttle, and at night my dad would take me outside to look at the sky. I was fifteen years old when Bluford became the first African American astronaut, and nineteen when he traveled to space. But until I worked at NASA Langley, it had never crossed my mind to become an astronaut, just like it had never been my goal to play in the NFL. And even at Langley, I didn't give it any serious consideration until my buddy Charlie got in. I had enough sense to know not to tell that to the interview committee as I responded to a range of questions, including queries about my experience handling tools, my manual dexterity, and how I faced the hardest times in my life. I had no idea how well I had done because all of my interrogators remained stone-faced and were careful not to give any indication of their thoughts.

After completing an interview round, the selection committee would meet with the twenty or so astronaut candidates who had been brought in that week. The "socials" took place at Petey's, a divey NASA hangout not far from Johnson Space Center. It was at Petey's that I got the opportunity to meet George Abbey, a shy engineer who spoke so softly you had to get closer to hear him, but who also happened to be the director of Johnson Space Center (JSC). It was widely known that getting a few minutes of face time with Abbey could help your chances. And, in a stroke of luck for me, he was a huge Dallas Cowboys fan.

John Young, generally a man of few words, was in a particularly sentimental mood when I saw him at Petey's. "Leland," I recall him saying, "once we stop exploring, as a civilization we will fail."

• • •

On a sunny morning in June 1998, I was making fiber sensors in the optics lab at NASA Langley when the call came. My team had just completed making the first optical fiber sensor, and everybody was feeling a great sense of accomplishment.

In astronaut culture, there are two calls worth celebrating: the one informing you you've been accepted into the Astronaut Corps and the one where you learn that you've been assigned a spot on a mission—that you're going to space. On that day in 1998, I knew that the astronaut selection board was starting to make the calls to applicants, and when I got to my office, I had a message from Ken Cockrell, call sign "Taco," chief of the Astronaut Office. I called him back, but the call got dropped. It happened again a second time. I thought to myself, *This is not a good sign.*

6

The World's Most Exclusive Club

dialed Ken Cockrell's number again and he picked up on the third ring.

"Leland, how's it going?" he asked me. Small talk. *Cockrell's letting me down easy, I'm sure of it.* "Huh, fine Ken. How are you?" He paused for what seemed like minutes. Now I knew he was messing with me. Finally, he said, "I'm calling to tell you that we'd like you to be part of the Astronaut Corps. Do you want to be part of the Astronaut Corps?"

"Yes, definitely. That's great," I said. I remember at that moment thinking to myself, *Wow, I'm an astronaut candidate. Okay, I'm ready. Let's go.*

But there was more. The second thing an astronaut candidate hears, after the invitation, are the words: "Don't tell *anybody.*" The NASA communications office wants to have a press release ready to go when the news breaks, and the communications office asks that you refrain from telling even your parents to prevent the news from leaking before it has a chance to control the announcement. By that point I knew NASA. So these instructions were no surprise. I sat there quietly, reveling in the moment on my own. That lasted about five minutes. "Hey, guys! I made it into the Astronaut Corps!" I called out to my team in the fiber optics lab. My ear was still warm from the call with Ken Cockrell. It just so happened I was accepted to the Corps on the very same day our team made its first optical fiber sensors, so there was already a sense of victory in the air. Then I called my dad. It was a good day all around.

Space travel remains one of the world's most exclusive clubs. Yet those long odds of becoming an astronaut have not deterred people from trying. NASA received more than 18,000 applications for its 2017 class, many attracted no doubt by the lure of Mars. The agency plans to choose eight to fourteen for space travel. Before that, the record for most applicants was in 1978, when 8,000 people applied as the space shuttle program was just getting off the ground. Now that the space shuttle has been retired, the opportunities to fly are a fraction of what they were back when there were six or seven launches a year.

The year I applied, for the class of 1998, twenty-five U.S. astronauts were chosen from a pool of 2,500 applicants with six international astronauts added to round us out to thirty-one. Most of those who have made it into the Astronaut Corps had been turned away before, some more than a dozen times. Peggy Whitson had applied and was rejected thirteen times before being selected into the class of 1996. She went on to become commander of the International Space Station, and served in that position during my first mission there. She later ran the Astronaut Office at the Johnson Space Center. But, before all that, she first had to fight to get into the Astronaut Corps.

Clayton Anderson, a 1998 Penguin, was accepted on his fifteenth try—fifteen years after he first applied in 1983. Equipped with a BS in physics and a master's in aerospace, Clay kept trying new things to get the attention of the board, like getting a scuba certification. In the end, he has no idea what worked, though he would tell everybody it had something to do with being from Nebraska, as the Corps had never had an astronaut from there. Mike Foreman was a naval aviator from my astronaut class who had logged more than five thousand hours on fifty different aircraft and became a flight instructor. He applied to the Astronaut Corps seven times before obtaining an interview in Houston. Still, it wasn't until his eighth attempt that he was accepted. NASA encourages applicants to keep trying. So they do.

Somehow the former football player with no smoldering aspiration to be an astronaut got in on his first try. The NFL Players Association calculates that the odds of a high school player getting into the NFL are about 0.2 percent. In 1998, I became the only person ever admitted in two of the most select clubs in the country.

● ● ●

How does one manage to succeed against such staggering odds? People ask me this all the time. The question is why does *anybody* succeed? Is it that people are born with a certain set of abilities, the right stuff, that predetermines they will someday fulfill their dreams? Or is it less about talent and more about attitude? A growing body of research shows that achieving big things is not reserved for those with certain God-given gifts. There are a lot more factors involved.

Carol Dweck, a Stanford psychologist, has produced some of the most intriguing thinking on success. She maintains that *what you believe* about your capabilities has a great deal to do with whether you will ever succeed. It all comes down to *mind-set*, she says. I think I have always fallen into the mind-set camp, before it was even labeled that. People with a growth mind-set believe their most basic abilities can be developed through hard work and dedication. This view creates an appreciation for learning that is essential for achieving anything great.

Dweck asks, "Do people with this mind-set believe that anyone can be anything, that anyone with proper motivation or education can become Einstein or Beethoven? No, but they believe that a person's true potential is unknown (and unknowable); that it's impossible to foresee what can be accomplished with years of passion, toil, and training."

Malcolm Gladwell offers an alternative to the raw talent argument in his book *Outliers: The Story of Success*. But he contends it's not all about mind-set either. Success, Gladwell says, has to do with historic circumstance and environment. And then there's Geoff Colvin, who believes talent is just one component that contributes to success. The author of *Talent Is Overrated*, Colvin says most great success is the result of something called deliberate practice and develops over time, often years and years. As an athlete, this is something I experienced. But it was the circumstances of my early life that also made my success possible. Guided and nurtured by our parents and inspired by pioneering role models in our community, my sister, Cathy, and I learned that we could do anything we wanted if we were willing to work for it.

• • •

It was early June when I got word of my acceptance into the Astronaut Corps, giving me nearly two months before I had to make the 1,400-mile drive southwest to Johnson Space Center in Houston. *Great*, I thought. Just enough time to finish up with my team at Langley and put my house in Hampton on the market. Joining the Astronaut Corps wasn't anything like getting just any new job; I was stepping into a new life.

My last afternoon at Langley happened to be the same day Franklin Chang Díaz was giving a speech there. Chang Díaz is a brilliant and respected Costa Rican–American astronaut who had already flown five shuttle missions and would eventually fly two more. He was something of a legend among legends in the space program. I asked him if he had any advice for me, the rookie. "Just be yourself," he told me. "Be authentic." His words stayed with me. I realized then that I was entering a world of public attention and celebrity that I hadn't fully understood until then. I'd received a certain amount of attention as an NFL draft pick, and I had seen what that kind of success could do to a player's character. I vowed I wouldn't let that happen to me.

Not long after, Woodrow Whitlow, at the time a fellow research scientist at Langley, threw a party for me at his house. A brilliant African American from Massachusetts Institute of Technology, Woodrow had relocated often as he made his way up the ranks. We followed each other around the country and would eventually wind up in Washington, DC, working as peers during my final four years at NASA. He was joined that night by Rosa Webster, Katherine Johnson, and many of the colleagues and technicians who had encouraged my growth. They were all part of a community that believed in me and helped me realize that becoming an astronaut was the right thing for me to do.

My final stop before heading to Houston was a celebration in Lynchburg, hosted by my sister, Cathy. It was an important step on my journey to come back to the community where I grew up, to the people who had helped me gain solid footing in the world. My neighbors and friends there, folks like Mrs. Williams, whose husband was the pastor of our church. The Powell family, our closest family friends, were there, too. They always made me feel supported and valued. My dad always got this certain look on his face when he was happy, and I remember he wore that look all day. As I hugged my mother, I knew that all my years of playing football had helped her prepare for the mix of fear and pride she was likely to feel in the years to come.

Astronaut Town

If you've never been to southern Texas, you don't know cockroaches. There's also unbearable heat, bad drivers, and racial prejudice that astonished even me. But the humongous size of the cockroaches was one of the first things I noticed when I settled into my new home in Houston. And the fact that they flew. *Flying cockroaches.* Let's just say Texas is not my kind of place.

I bought a house in El Lago, the same subdivision where

Neil Armstrong and Buzz Aldrin lived at the time of the first moon landing, and where Jim Lovell called home during the treacherous *Apollo 13* mission. If you've seen the movie *Apollo 13*, El Lago is aptly portrayed as the cookie-cutter neighborhood where Marilyn Lovell and the other astronaut wives gathered as their husbands faced an uncertain return to Earth. El Lago City Hall has an Astronaut Wall of Fame with photos of all the astronauts who had lived there—forty-eight at last count, including me. Buttressed by Taylor Lake to the west and Clear Lake to the south, most of the residences in El Lago back then were modest four-bedroom ranch houses surrounded by big lawns, though in recent years ornate show-pieces have replaced many of the modest homes from those days. The place I found was a simple yet beautiful one-story house with a bougainvillea that crawled up the wall of an atrium. I remember thinking, *I could get used to this.*

On the other hand, some people had to get used to me. Let's just say there were probably fewer African Americans in El Lago when I arrived than there were black astronauts in NASA, and that's not saying much. I'll never forget the day I moved in. A woman across the street stared at me, her arms folded across her chest. "Hi!" I said and waved to her. But she just shook her head and walked into her house. Thanks for the warm welcome, neighbor.

Becoming a Penguin

"Men wanted for hazardous journey. Small wages, bitter cold, long months of complete darkness, constant danger. Honor and recognition if successful."

Legend has it that the explorer Sir Ernest Shackleton ran this ad in the *Times* of London when he set out to find the perfect crew for his 1913 expedition to the South Pole. He got five thousand responses for twenty-eight positions and had to turn away hundreds of qualified candidates. I'm quite certain that all the men and women in the 1998 class would have answered that ad.

It's a tradition in the astronaut program for the previous class to bestow a name on the new class. Our predecessors, the Sardines, chose to call us the Dodos, after the flightless bird. (They said it would be a while before any of us flew, if ever.) Not eager to be associated with an animal that became extinct and happened to be quite ugly, we changed our name to the Penguins. We were still identified with a flightless bird, but at least a more elegant one.

On my first night in town, we met up for a reception at the home of Gregory Johnson, a retired Navy captain who went by his pilot handle, Ray J. As the oldest member of the class and a veteran of Johnson Space Center since 1990, Ray J became our class leader, along with army helicopter pilot Timothy "TJ" Creamer. TJ was an army aviator who had been working in the space shuttle program for the past three years. Clearly NASA brass thought the Penguins needed some leadership with inside knowledge of the organization, and I'm sure they were right.

The Penguin class brought together twenty-five decorated military-trained pilots, research scientists, and a school-teacher to "pull us into the future," as Ronald Reagan would say of space travelers. (Another six candidates were joining us from other parts of the world, but they weren't scheduled to show up in Houston for another month.) Decades of scientific research and trial-and-error devoted to what makes a good astronaut had brought us to that point, but looking around the room you couldn't have guessed what we had in common. Still, from that moment forward, everything we did was designed to solidify us as a team.

Many years later I would realize just how much my training in football helped me be a better astronaut. Both require the same notion of teamwork: Just like the quarterback sets the tone and tells the team what play to run, the commander of a shuttle mission might say, "We've got a malfunction and it's time for you to take on this role." Like the quarterback and the wide receiver, the shuttle pilot and the mission specialist develop a synchronicity—they can anticipate each other's next move.

Through football I also acquired a perfectionist mind-set and concentrated on being exact and deliberate. I learned to control the adrenaline rush and stay focused. Space travel involves the same kind of control of your acceleration. In football you propel yourself from a dead stop toward an opening, somewhat similar to pushing yourself off the walls of the space shuttle in the zero-gravity environment of space. In both cases, success can hinge on decisions, by multiple players, that take place in an instant. This was all part of the training I embarked on that first year.

• • •

On a bright Texas morning in August 1998, I walked onto the hallowed grounds of Johnson Space Center eager to make my mark on the program and advance civilization by boldly exploring new worlds. I also remember wondering when they were going to serve lunch.

Because I never had a burning desire to become an astronaut, I hadn't spent my teenage years avidly following the career trajectories of John Glenn and Alan Shepard and wondering what I had to do to be just like them. At least I didn't before my friend Charlie applied and flew into NASA Langley on a T-38. At the time, I didn't know much about John Young, except that he'd once walked on the moon. But even that knowledge didn't make a big impression on me. So it didn't bother me too much when on that first day at the Johnson Space Center I was reminded I was just an "AsCan," short for astronaut candidate and pronounced "ass-can." The label pretty much says it all. Until you've finished training, you're just a rookie. You haven't proved a thing. For many astronaut candidates, this is the first time they haven't been the smartest guy in the room. I'd been a journeyman plenty of times, the guy who got things done. It reminded me of my experience in the NFL, except this path, I hoped, had a future.

Before too long I would find myself steeped in the astronaut world. NASA's culture of perfection is a humbling experience at first, but one that reinforces the awesome responsibility that is part of becoming a space traveler. Consider that NASA invests tens of millions of dollars to train a single astronaut. And these days, until we have an alternative, the United States pays Russia at least $80 million for each seat on a *Soyuz* rocket to taxi astronauts to the International Space Station. It wasn't until that first day at the space center in Houston that it hit me. I was joining this elite club. Me, the unexpected astronaut.

• • •

The first year of training involves moving from one activity to another like kids do at summer camp. Boats! Planes! Swimming lessons! Plus there's robotics, extravehicular activity training, and survival training in water and on land, plus time (lots of time) in the space shuttle simulator.

But it was the jet that got me. The T-38 is NASA's two-seat, twin-engine, supersonic flying machine that has been a fixture at the space center since the 1980s. It trains you to think fast and adapt to changing situations—while wrenching through seven G's—seven times the force of gravity—while performing acrobatics. Operationally, it's one of the closest things to flying in the space shuttle you can experience. Even a non-pilot mission specialist like me had to get hours in the jet because NASA's research shows that skills like "cockpit resource management" benefit a crew working in a highly dynamic environment like flying in space. All the navigation skills learned in the jet can be used aboard the space shuttle, the Russian *Soyuz*, or operating the controls of the International Space Station. It was imperative that we all "got our mins," our quarterly minimum flight time in the jet, to demonstrate operational safety and proficiency.

I had some of my best moments in the backseat of a NASA T-38. One time I was sitting low in the backseat when I caught a glimpse of Mike Anderson's face in the rearview mirror, his broad smile indicating complete happiness. We were making our slow ascent from Cape Canaveral after watching the space shuttle *Discovery*, STS-96, soar into the cloudy sky.

Mike was one of only two other African American astronauts at NASA when I arrived and he was the only "front-seater," meaning he had military-trained pilot status and could fly the T-38s. Before I arrived at Johnson Space Center, Mike had taken his first shuttle flight aboard *Endeavor*, delivering equipment and fresh water to the Mir space station. A few years later, he would never return from his second flight, as payload commander aboard *Columbia*.

When we took to the air on that beautiful morning in May, the sky was full of the kind of bright billowy clouds that blow in after a Florida thunderstorm. We'd been cleared to ascend to 40,000 feet, but Mike had no intention of taking the direct route. Rather, he steered the jet toward the sun and banked through the clouds like he was gliding through fresh snow. It was perfect.

Many years later, while watching the movie *The Tuskegee Airmen*, I was reminded of that day in the sky with Mike. There's a scene where one of the airmen, played by Malcolm-Jamal Warner, is piloting a sputtering plane, searching the landscape for a place to make an emergency landing. Somehow he and his fellow pilot (played by Laurence Fishburne) manage to softly guide their planes onto a country road, where two white prison guards with shotguns are overseeing an all-black crew of prisoners clearing brush. When the pilots climb out, happy to be alive, their oxygen masks are obscuring their complexions. The white guard is fixated by what he just witnessed and eager to meet the heroes. "They's our boys," he says proudly. Jamal-Warner and Fishburne take off their masks, revealing their faces. *Black pilots?* The guard sputters in confusion while the camera zooms in on the proud faces of the prisoners. I felt similar pride when Mike and I sailed through the clouds over coastal Florida.

Mike loved cars almost as much as jets, and he particularly loved his gray Porsche 911. I remember the sound of that car's engine as Mike whipped it around the corner and into his parking space at the space center. The Porsche made a steady, comforting hum, and when we heard it we knew Mike had arrived. But the car was too small for his growing family—he had two daughters—and, reluctantly, he decided to sell it. We arranged for me to buy it after his *Columbia* mission, agreeing that he would still be able to drive it whenever he wanted. But after the *Columbia* accident, there was no question that Mike's wife, Sandra, would want to keep the car for the girls, and I found another Porsche 911 to buy. For years after, the hum of the engine around those same sharp curves made me think of my old friend.

Another time I was getting my "mins" with Jim Wetherbee, a Navy aviator who had already flown four shuttle missions, including one as pilot and three as commander. He would later fly two more. As director of flight crew operations, Jim was tough on AsCans. "You need to fly more!" he would tell us, again and again. No one wanted to fly with him because you knew that if you made a mistake or didn't stay "ahead of the plane" it could affect your getting assigned a spaceflight. At least that is what some AsCans thought.

Being "ahead of the plane" means staying on top of all the checks and details under your responsibility, like calculating the fuel and moving the radio dial to the next frequency you're going to use. You try to do as many of these things ahead of time as you can. By developing this "situational awareness" you are training yourself to identify potential problems before they arise. When you're flying with a guy like Wetherbee your every move is carefully observed and evaluated. The idea is if you can do it in the jet, you can manage in the space shuttle, which means the reverse assumption is also true—if you make a mistake in the jet, they assume you could make it in space. So most AsCans simply avoided flying with pilots like Wetherbee.

But there I was, in the backseat of Wetherbee's plane. I was in charge of setting the altitude alerter, which tells you when you're nearing a certain altitude range. If you were out of your intended range, the melodic recorded voice of Bitching Betty would come on the speaker and issue a warning.

On that day, Wetherbee and I got word from air traffic control that we were "cleared for flight level 1-8-0," after visual confirmation that there were no aircraft in range above us, and we could ascend to 18,000 feet. I dialed in 1-8-0 and informed Wetherbee, who replied, "I see it." I took that to mean that he saw the traffic above, when in fact he was merely confirming that he'd seen me dial in our target altitude. We got lucky that day because there were no aircraft in our path, and I'm sure I became a better navigator after that.

• • •

Near the end of our first year as AsCans, I looked out the window as my plane made its final approach over Brunswick, Maine. I noted with interest the dense forest and rugged peaks below. *Looks cold*, I remember thinking. We were about to endure four days of survival training, like Outward Bound for astronauts—but with an entourage. Paige Maultsby, our "den mother," Duane Ross (head of astronaut selection) and Robert "Beamer" Curbeam Jr. had accompanied us. Beamer had completed his first mission aboard *Discovery* and had been assigned to shepherd us through these exercises. We were also all under constant observation by a few psychologists to help assess our attitude and our skills. Ours was only the second astronaut class to be evaluated for both short and long-duration missions, which meant we really had to show we could get along. We also had to prove we could be leaders, as well as leaders who could follow if the circumstances demanded it.

For the land survival part of the exercise, I was paired with Sunita Williams, a Navy test pilot and triathlete. I would quickly discover that she was the toughest and most competent partner you'd ever want. On a two-month stint as commander of the International Space Station in 2012, Sunita would become the first person to ever complete a triathlon *in space*. The hardcore military aviator and the civilian former football player were an odd couple, but we formed a bond. We were instructed on navigating, reading maps, constructing shelters, finding our own food (not to mention skinning it), and building fires. A driving rain on the second day kept us from a good night's sleep, but other than that it wasn't too bad. Except for the part with the rabbit.

Sitting around the makeshift campground in a clearing, we watched as the team leader reached into a box, pulled out a jackrabbit, and quickly dispatched it with a whack on the back of the head. Then he proceeded to skin it, pausing now and then to describe the nutritional benefits of a certain organ. That was bad enough. But he asked for a volunteer to eat the rabbit's *eyeball*, explaining it was a good source of salt and protein that could help keep you alive if your vehicle went down. Somebody cracked a joke about there probably not being many rabbits in space as Clayton Anderson, a fellow Penguin who was also eager to impress his fellow AsCans, reached for the slimy ball and popped it in his mouth.

Star City

As my first year of astronaut training was winding down, NASA leadership began talking about what branch the astronaut candidates would be assigned to and where they would specialize. Each candidate is asked to write down the top three choices of jobs they'd like to have. It was common knowledge that some of us would be needed in Russia, where the three-person crew of Expedition 1 was deep in training to become the first long-duration occupants of the brand new International Space Station. The crew would fly on a Russian *Soyuz* rocket. Liftoff had been scheduled for October 31, 2000, at the Baikonur Cosmodrome in Kazakhstan, which is 1,600 miles southeast of Moscow. The Cosmodrome was the same site from which Yuri Gagarin in 1961 became the first human to fly in space. However, before the Expedition 1 launch could take place, a million things needed to happen and most of them in Moscow.

The Expedition 1 crew consisted of two Russian cosmonauts, Yuri Gidzenko, a lieutenant colonel in the Russian Air Force, and flight engineer Sergei Krikalev. The sole American crewmember, Bill Shepherd, would be the commander of the International Space Station and only the second American launched in a *Soyuz* capsule, following on the heels of Norm Thagard, who in 1995 became the first American to live and work aboard Russia's Mir space station.

Today's cooperation in space exploration between Russia and the United States is a far cry from the days of the Cold War. Back in the mid-1950s and early 1960s, the two nations were rival world powers. Each nation sought to prove its superiority, from athletic competition and educational standards to military strength and new technology. Space exploration was part of the mix.

On October 4, 1957, the Soviet Union launched Sputnik, man's first satellite, into space. The launch marked the start of the space race as America scrambled to stay competitive in what many saw as a new front of the Cold War. In 1958, the U.S. launched its Explorer 1 satellite, and President Dwight Eisenhower signed an executive order creating the National Aeronautics and Space Administration.

The space race heated up in 1959 when the Soviets successfully launched the Luna 2 probe to the moon. Two years later, Soviet cosmonaut Yuri Gagarin became the first person to orbit the Earth. Alan B. Shepard Jr. became the first American in space but not in orbit. John Glenn would get that distinction in 1962. Still, the progress with Shepard and the new space agency prompted President Kennedy to boldly claim that America would put a man on the moon within a decade. When that finally happened in 1969, the U.S. claimed victory. In 1975, a joint *Apollo-Soyuz* mission sent three U.S. astronauts into space to dock with the Soviet-made *Soyuz* craft. When the two commanders greeted each other high above the Earth, their "handshake in space" symbolized the end of the space race era.

As one of the few unmarried astronaut candidates in the Penguin class, I knew my NASA bosses would see me as a prime candidate for working with the International Space Station branch in Russia. But it was not without risk. Spending a year abroad in the role of crew support, rather than working at Johnson Space Center under the nose of the top brass, could delay my chances at going to space.

As my first choice of branches, I requested extravehicular

activity, which I figured would give me a jump on spacewalk training and, with some luck, hasten my assignment to a space shuttle mission. It was the most sought after assignment, one that rarely went to a rookie. I wrote down the International Space Station branch as my second choice, knowing I was probably on the short list for the space station anyway. The robotics branch came in third. I wasn't surprised when I got a call from Stephen Oswald, in the Astronaut Office at Johnson, telling me I'd been assigned to go to Moscow to be what they liked to call a "Russian Crusader." From the moment I landed in Moscow on a warm summer day in 1999, I felt as though I was embarking on an adventure.

Moscow itself is a beautiful, vibrant, but complex city. It reminded me a little bit of Chicago, a metropolitan area with a little bit of everything—from downtown tourist attractions and trendy neighborhood hotspots to rundown areas plagued by crime and poverty.

Every time I visited Red Square, I'd think of the film *The Saint*. The famed city square that separates the Kremlin, the old royal citadel, and the official residence of the Russian president is immaculate and typically filled with Russians and tourists. The mosaic tiles in the city's subway system are intricate and tell the history of the country. There's plenty of propaganda in the stories left over from the old Soviet era, but the tile artwork lining the subway system is breathtaking.

It didn't take long though to realize that I wasn't in America. People on the street don't immediately warm up to strangers. They are suspicious, which can be hard for many outgoing American travelers. I got a lot of stares while walking the streets. Sometimes, curious children would come up to me and say hello, and it wouldn't be long before their parents would pull them away from me. I attributed a lot of that to being an African American in a largely white city. Moscow is one of the world's most ethnically diverse communities. Yet black skin remains rare.

Once I settled into a neat apartment at the Volga Hotel, I began to think about my training. A van took me from the hotel to Energia, the Russian government contractor working on the space station and the *Soyuz* rocket. I was assigned to work with Russian interpreters in translating the procedures for the *Soyuz* and the space station. These interpreters were competent and highly trained, and some of them had considerable technical backgrounds. But that wasn't enough. These Space Operation Data Files (SODF), as they were called, contained the procedures that astronauts on the space station would follow. My job, and that of my fellow crusaders in Moscow, was to verify those translated instructions and to make sure they worked by testing them in a computer simulation and in a simulated space capsule. The Russian space agency knew the instructions to be accurate and effective in Russian. Yet, even a subtle mistake in translation could be disastrous, or in one particular case, humorous. I received a rather profane Russian-to-English translation to describe the use of a torque wrench to determine how much force was needed to attach a bolt. The translated procedure read "F—k the bolt to the hardstop." It should have read "Screw the bolt to the hardstop." I sheepishly approached the translator with the correct word. It was one of the funnier moments over there.

However, the assignment at Energia didn't last long. About a month after I arrived, I got a call from Houston. It was Anna Fisher, chief of the space station branch. Fisher had been an emergency room physician in Los Angeles before joining the

Astronaut Corps and logging 192 hours in space. "Leland," she said, "we need you to go to Star City and help take care of Shep. He's going to be the first commander of the International Space Station . . . but there are some problems."

Shepherd had spent almost two years living and training at Star City, the compound fifteen miles outside Moscow that has served as Russia's cosmonaut training facility since the 1960s. Nestled deep in a birch forest, the official name of the compound called Star City is the Gagarin Cosmonaut Training Center, named for the U.S.S.R.'s Cold War hero. During the Soviet era, it had been a secret, heavily guarded military installation and it still felt a lot like one when I arrived.

Shep's bulldozer ways hadn't been winning him any friends at Star City, among either the Americans stationed there or the Russians, and NASA desperately needed someone to serve as a buffer. Shep could be controlling and sometimes stubborn and the top brass at Johnson Space Center was frustrated with his need to be involved in even the smallest details of the upcoming launch. Ginger Kerrick, a NASA instructor, was having some success trying to contain him, but Shep needed someone specifically assigned to getting him through this final stretch.

Inside NASA, if you knew only one thing about Bill Shepherd it was probably this: During his interview for the Astronaut Corps in 1984, the former Navy SEAL was asked what special skills he had. He reportedly responded, "I know how to kill somebody with a knife." That reply soon became part of NASA folklore and Shepherd himself became something of a legend. Even today, if a Navy SEAL makes it to interview level, someone will invariably ask with a smirk, "Can you kill with a knife too?"

I packed up what little I'd brought with me to Moscow and moved in with Shep in Cottage Number 3 at Star City. The assignment would last fifteen months and I flew back and forth from Moscow to Houston supporting Shep in both locations as needed, until a *Soyuz* rocket safely transported Shep, Gidzenko, and Krikalev to the space station.

I became Bill Shepherd's body man. Our cottage was one in a cluster of two-story duplexes for Americans, each with three bedrooms upstairs and a living area on the first floor. The house also had a basement with a weight room, as crude as it was, and an area set up for video conferencing with the folks back at Johnson Space Center. We also had our own private built-in bar—Shep's Bar, we called it, installed specifically for him. I soon learned the merits of having a bar in your basement.

My job was to accompany Shep on his training exercises, to the massive hangars that contained the *Soyuz* simulators, and wherever else he wanted to go—anywhere. That sometimes included brisk, cold walks to a lake on the grounds of Star City where Shep would share what he knew about drawing and painting while I taught him the principles of photography. I had loved photography since middle school and took my camera along with me everywhere, a hobby that would serve me well when I flew to space. During that time I got to know Shep and that his concerns were really aimed at ensuring the crew would operate safely. As a Navy SEAL he was trained to work as a self-contained operator and his actions and preparedness would determine the outcome, life or death. So I started to understand why he wanted to know all the details, because he was ultimately in charge of this new multibillion-dollar outpost as the commander.

My obsession with photography that year, coupled with my need to get away from Star City every now and then, took me all over Moscow. I felt relatively safe, despite getting stares wherever I went. The Soviet era had ended eight years before and the city of Moscow was quickly becoming a tourist destination for western Europeans and Americans. Yet I almost never saw another black person. Every now and then I would meet a college student from the African country of Cameroon, which had an exchange program with Moscow universities dating back a few years, but that wasn't often.

I wasn't unaware of the city's potential dangers. Russian police can confiscate your passport, then leave you in a place where their buddies could rob you. Gypsy taxi drivers can drive you off somewhere, rob, and kill you. The year before I arrived, a gang of skinheads had beaten up a black Marine, knocking out two of his teeth. We had been warned to take caution, but I liked walking around by myself, trying to absorb the language and culture.

One Saturday afternoon, I set out for the outdoor market in Filevsky Park, one of the city's greenways along the Moscow River. After I passed through a security gate, I headed for the back of the market where it was quieter, when I noticed about a dozen skinheads staring me down. The situation grew increasingly threatening, and soon they made it clear they had plans for me. I needed to get out of there. I knew NASA would not be happy if I showed up in a Moscow hospital. I turned and ran for the exit, which seemed to be just the incentive they needed. They were in full pursuit. My speed kicked in and my hamstring didn't fail me. I finally shook them. Cowboys quarterback Danny White would have been proud.

Our family portrait when I was two years old, May 1966. *(Courtesy of Melvin Family)*

Number 47 on the Perrymont Elementary Panthers football team. *(Courtesy of Melvin Family)*

In my freshman year at Heritage High School. *(Courtesy of Melvin Family)*

Making the winning catch during the Rustburg versus Heritage High School homecoming game. *(Courtesy of Heritage High School)*

Teaching City Recreation Tennis with my friends Addison Fauber and Kimbrough Richards, at Heritage High School, summer 1984. *(Courtesy of Melvin Family)*

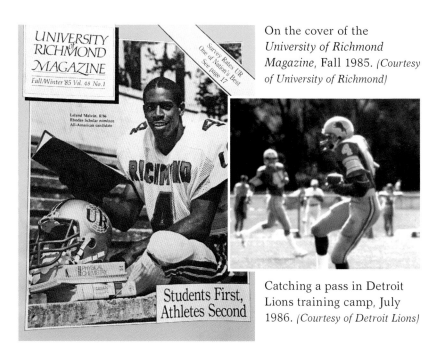

On the cover of the *University of Richmond Magazine*, Fall 1985. *(Courtesy of University of Richmond)*

Catching a pass in Detroit Lions training camp, July 1986. *(Courtesy of Detroit Lions)*

Group 17 Astronaut Class, Johnson Space Center, Houston, Texas, August 1998. *(Courtesy of NASA)*

The "Big Blue Crew" at a celebration honoring me as the first astronaut from Lynchburg, summer 1998. *From left:* Bryant "Boogie Bear" Anderson; Phil "Silky Blue" Scott; me, "Lil D"; Ernest "Fufu" Penn; and Kip "Gus" Hawkings.
(Courtesy of Melvin Family)

With my hearing specialists, Nancy and Dr. Bobbie Alford, during an examination, April 2001.
(Courtesy of Melvin Family)

With Jake and Scout after a run in Ronald McNair Memorial Park, El Lago, Texas. *(Courtesy of Lynchburg News Advance)*

Jake and Scout on the road again, going on another journey. *(Courtesy of Melvin Family)*

Driving an armored personnel carrier, with Alan "Dex" Poindexter *(far left)* looking on, fall 2007. *(Courtesy of NASA)*

Me in my Detroit Lions jersey on the flight deck of the space shuttle *Atlantis* during mission STS-122, February 2008. *(Courtesy of NASA)*

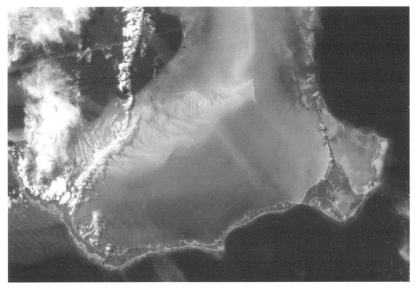

The view from spacc of Rigel Island, in the beautiful Caribbean. *(Courtesy of NASA)*

My big sis, Cathy, was waiting with a loving embrace when I returned from Ellington Field, Houston, Texas, February 2008. *(Courtesy of NASA HQ)*

The mission STS-122 welcome-home ceremony at Ellington Field. *From left:* Astronauts Dan Tani, Stan Love, Hans Schlegel, Rex J. Walheim, me, Alan Poindexter, Steve Frick, and Center Director Michael Coates. *(Courtesy of NASA)*

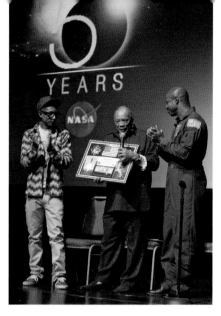

Pharrell Williams and I present the music legend Quincy Jones with my space montage at a ceremony in Washington, DC, 2008. *(Courtesy of NASA)*

The crew of STS-129 as we walk out from crew quarters to board the Astrovan to head to the launchpad, November 2009. *Front:* Charles "Scorch" Hobaugh and Butch Wilmore; *middle:* me and Randy "Komrade" Bresnick; *rear:* Mike Foreman, Bobby Satcher, and Peggy Whitson. *(Courtesy of NASA)*

Atlantis liftoff from Cape Canaveral, Florida, November 16, 2009. *(Courtesy of NASA)*

On the International Space Station before heading to *Atlantis* to prepare for our trip home, November 2009. *Front row:* me, Nicole Stott, Bobby Satcher. *Back row:* Mike Foreman, Charles Hobaugh, Butch Wilmore, and Randy Bresnick. *(Courtesy of NASA)*

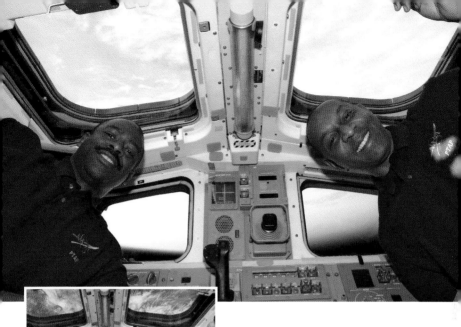

Above: Me and Bobby Satcher, making history: the first time two African American men flew together in space, November 2009. *(Courtesy of NASA)*

Left: Surrounded by 2,700 calories—one day's worth of food in space, November 2009. *(Credit: NASA)*

An image taken from the space shuttle *Atlantis* as we fly around the International Space Station, November 2009. *(Courtesy of NASA)*

Sitting among a sea of inspired kids, Cape Town, South Africa, October 2011. *(Courtesy of NASA)*

Showing much respect for my friend Astronaut Mike Anderson at the Seattle Aviation Museum, Washington. *(Courtesy of Lonnie McCool)*

The NASA education team with a true legend, Katherine Johnson, at NASA Langley, Hampton, Virginia, 2013. *(Courtesy of NASA)*

At the Smithsonian Air and Space Museum with NASA astronaut candidates and, in the background, International Space Station Astronauts Rick Mastracchio and Michael Hopkins, calling in from space. *(Courtesy of NASA)*

With President Barack Obama and First Lady Michelle Obama during the White House Easter Egg Roll, April 2014. *(Courtesy of White House)*

As host of the Lifetime TV show *Child Genius: Battle of the Brightest*, 2016.
(Courtesy of Lifetime Television)

Speaking at a Black History Month event at NASA headquarters, February 2016.
(Courtesy of NASA)

Katherine Johnson receives her Silver Snoopy Award from me at NASA Langley, September 2016. *(Courtesy of NASA)*

On the red carpet of the *Hidden Figures* movie premiere in New York, November 2016. *(Courtesy of NASA)*

Astronauts Ron Garan and Anousheh Ansari, Bill Nye, and me at the Global Citizen Festival in New York City, 2016. *(Courtesy of Melvin Family)*

We had been trained to avoid fixating on problems, to com-partmentalize dangers, and make sure that we didn't repeat a mistake again. The important thing was moving on to the next thing, and that's what I did. I was determined not to let anyone else—including a gang of crazed skinheads—influence my thoughts and actions to the point where I could not do what I needed to do.

Star City had no real nightlife, and that usually wasn't a problem. On this one particular night, however, we decided to break the monotony by throwing a party in our private bar. That was our first mistake.

Shep, Ken Bowersox, who was crew backup for Shepherd, and Bob Cabana, manager of international operations from the Johnson Space Center, were there, along with Ginger Kerrick and me. To get the party going, we started playing a military drinking game called Liar's Dice. A handful of astro-nauts blowing off steam in the basement of a high-security compound miles from the city. *What kind of trouble could we get into?* The evening would have been fine enough with the camaraderie and the drinking game, but after a few rounds Bob and "Sox" decided to call our boss in Houston, George Abbey, the head of Johnson Space Center.

I remember Bob reaching Abbey and saying a few words. He then handed the phone to me. I sobered up enough to get out a few words about what an honor it was to be of service in Moscow, followed by a quick "Goodbye, sir."

To an astronaut candidate in the space program, George Abbey was the single most powerful—and feared—person at NASA. He had the final say in making mission assignments, and unlike Cabana, who had already flown on four shuttle missions, I hadn't been assigned to my first mission.

The hijinks, however, didn't stop there. Abbey, still on the line, asked Cabana to hold and transferred our call to Dan Goldin at NASA headquarters in Washington, DC. Goldin was no ordinary manager. He was NASA administrator; he was in charge of America's entire space program. Once Bob realized who was on the other end of the line, he sobered up, fast.

What happened next depends on who's telling the tale. Bob remembers both Abbey and Goldin taking the call well. Abbey thought we were bonding, a good sign. Goldin, he said, "loved it." Ginger Kerrick, the only one of our group who did not partake in the drinking game, may have had the more accurate assessment. "As for Goldin, oh heck no, he was not happy," she said. "While Abbey [initially] did take it well, when he transferred the call to Goldin he was upset. That's the whole reason I unplugged the phone."

The day of the Expedition 1 launch was fast approaching. Shep, Yuri Sergi, Ginger, and I flew out of Star City in a plane that carried Russian troops to Kazakhstan a few days before the launch. When we landed in Baikonur, Kazakhstan, I saw a few things that I didn't expect. Let's start with the camels, a major form of transportation. The people appeared to be out of place, too. Many looked like Asians, which seems odd when you think of Russians. Of course, it's easy to forget Kazakhstan borders Russia and China.

Baikonur sits on the Syr Darya River. The land is barren, dry, dusty, and cold. The town itself was originally built to service the Baikonur Cosmodrome, where Yuri Gagarin made history as the first man in space. The launch site occupies a vast dry plain in the middle of nowhere. The town probably hasn't changed much since Gagarin's flight. Our quarters had a drab, muted yellow institutional look from a bygone era, but we made do.

As the crew milled about, Ginger and I set out to explore the town. I loved to cook, and I had heard you could buy saffron at a cheap price in this part of the world. I wanted to make a paella, with sausage, rice, chicken, peppers, onions, and deep golden saffron. The markets were full of people bartering their wares of beef, cloth, rice, and saffron. There's not much to do in Baikonur. No one was in a hurry, as compared with people living in Houston, New York, and other big American cities. I bought a pound of the spice for twenty-five rubles, about one dollar. Back home, you'd pay a king's ransom for one-tenth of that amount.

The town had another surprise. We passed by an open area that contained an old buran—a Russian version of the space shuttle. The relic sat outside, having weathered many years since the Russians tried to copy our space shuttle program. This buran flew once, but the Russians decided to forgo further development when the American and Russian space programs opted to build the International Space Station. The vehicle was big and angular like much of the Russian architecture. You wondered how something so big could even fly.

While driving to the *Soyuz* launch area, we passed a park where kids were swinging, and as soon as they saw the bus, they lined up military-style, looking in awe and wondering what Russian military generals were leading this crew. Historically Russian commanders headed the Mir space station, but not this time. The kids didn't know they had saluted the first American commander of the International Space Station, flanked by his two Russian engineers.

I could only imagine the excitement building up inside Shep. He had flown in space three times before, all on shuttle missions that lasted about a week. This one would be very different. He and his crew would travel in a Russian-built *Soyuz* TM-31 spacecraft to the space station, making him the first American to undertake a long-duration flight. It would be the start of uninterrupted human presence in space.

The Russians forgo the frills typically associated with a NASA launch, but that didn't dissuade Shep. He was in good form and eager to get on with what he had been preparing for the past three years: a nearly four-and-a-half-month stay on the International Space Station as commander of the outpost's first crew.

The next time I saw Shep, it was on Christmas Day when I patched a call from him on the space station to his family in the States. A few days later, I was back home in El Lago, Texas.

The Prophecy

As an astronaut training to fly in space you always took note of airport runway lengths because if you had 7,500 feet of asphalt you could get a pilot buddy to fly you wherever you needed to go in a NASA T-38. But on this spring day I was flying commercial because my hometown airport in Lynchburg only had 5,500 feet of tarmac.

It had been nearly four months since I'd returned to Houston from Moscow and I had four days before I was to report for extravehicular activity training. I was heading home for my parents' thirty-fifth wedding anniversary. As soon as I landed, my duties were to help the guests get settled from their travels so that on Sunday they could witness Deems and Gracie Melvin's celebration of marital bliss. Despite teaching at the same school, my parents never tired of each other's company. They drove to work together, bumped into each other in the hall, and then drove home together—every day. They were happy and truly loved and respected each other in a very powerful and amazing way. This celebration was another example of my parents' ability to bring people together.

Not long after arriving in Lynchburg, I drove to pick up my cousin Phyllis McLymore and an unexpected passenger, Jeannette Williamson Suarez, to take them to a hotel downtown. But as soon as we got there, a sudden downpour made it nearly impossible to get out of the car. So we waited. They asked about my life as an astronaut, and we talked about my parents and the upcoming celebration. Then, out of the blue with the fire and brimstone of a Southern Baptist preacher, Jeannette called my name and said she needed to share something with me. I had just met Jeannette and did not know she was a minister near the small town of Roseboro, North Carolina, where my father had grown up.

I turned around, looking at Jeannette in the backseat, and she looked me in the eye. "I have a prophecy for you," she said and told me that something unexpected would happen to me, and experts around the world would not understand why it happened. Yet I would overcome this setback, fly in space, and this would be "my testimony to share with the world."

Now this was a heavy message to hear. I had just returned from Russia where I worked with the Expedition 1 crew. I had developed situational awareness in a jet and mastered space shuttle and station training. I learned how to scuba dive and had dived in the Red Sea near Sharm El Sheik, Egypt, in an area called the "washing machine" because the currents were so strong you could be taken for a ride seventy-five feet under the surface. I had endured Navy water and land survival training and in just four days I was going to start my extravehicular activity training in the six-million-gallon Neutral Buoyancy Laboratory to see if I had "the right stuff" to walk in space.

Still, I listened respectfully and humbly to Jeannette. She had a divine nature about her, and I knew her words were not those of a charlatan but had a seriousness and truth to their delivery, and a relevance to my journey. I had never had someone share a prophecy with me before, unless you count the time at a county fair when someone read my palm and shared generic things that I already knew. In retrospect, for someone to predict something so specific would normally have unnerved me. I thanked her for her words as the spring showers dissipated and the sun started to peek out from the clouds, and we set off again.

Four days later, I put on the puffy white extravehicular activity suit and climbed into the pool at the Neutral Buoyancy Lab, ready to start my journey as a future astronaut.

7

Recovery and Tragedy

Jeannette's prophecy turned out to be an accurate description of my ordeal: the descent in the pool, the static in my headset, the sudden silence, the hapless investigations of concerned doctors and colleagues. I spent the next few weeks continuing my recovery at home in El Lago. I slowly began to regain hearing in my right ear but not much in my left. I struggled to understand my place in the world. Some days I was overcome with despair. With still no explanation for my sudden hearing loss, it was looking more certain than ever that I would never fly on a mission to space.

During my second week at home I got a visit from Eileen Collins. She wasn't in my class but she made it a point to come by and see me. Collins was NASA's first female space shuttle commander. The only other woman to do that was Pamela

Melroy in 2007. Collins would return to space in 2005 as commander of the Return to Flight shuttle mission after the *Columbia* tragedy. She lived much of her life in the spotlight, but she was very warm and humble. "It's going to be okay," she said to me, with the confidence of an Air Force test pilot and the warmth of a mother. "If there's anything that you need, I'm there for you."

Eileen was one of my few visitors during this period, possibly because the Astronaut Office had advised people to avoid hindering my recovery. The solitude was fine by me, because I preferred to be alone while my brain was trying to rewire itself to hear again. I've always been an introvert, and my impaired hearing made me retreat to an even deeper solitary place. I was trying to figure out what my life in the Astronaut Corps would look like. I questioned many things but kept coming back to Jeannette's prophecy and wondered how that was going to manifest in my life. It seemed that sometimes they were just words. Did I lose my faith? "Lost" may be too strong a description, but I did question it. Did God really want me to fly in space? Is there a God? What's out there? Perhaps I'd have to figure out the answers with my feet on the ground. Space, the doctors said, wouldn't be happening for me on their watch.

In the midst of coping with my circumstances, I learned that Patty Hilliard Robertson, a member of the Penguin class and a close friend, had been critically injured during the crash of a small plane in Manville, Texas, and she was fighting for her life. Her husband, Scott, was asking that I come to the hospital. It wasn't until I arrived at Houston Methodist that I grasped the tragedy that was unfolding from the anguished faces of Patty's family. She was in a coma and not expected to survive.

Patty had been a passenger in a two-seat experimental aircraft performing touch-and-go landings when the left wing clipped the ground, sending the plane into cartwheels down the runway before crashing into a tree. Now Patty was clinging to life in the same hospital where I had spent much of the previous month.

Only three weeks earlier, Patty had waltzed into my hospital room with a plateful of sushi. She loved food almost as much as I did. Early in training, Patty and I had confessed to each other our mutual love of food, and during downtime we would talk about what we planned to make for dinner that night, a new restaurant in town, or what might be on the menu during an event coming up. Her mom, Ilse, was from Indonesia and really knew how to throw down on some good, spicy dishes. When Patty married Scott, she would ask me relationship questions to learn more about the male perspective. A classic NASA overachiever, Patty was a doctor who had left a thriving pediatric practice to fulfill her lifelong obsession with flying. Like me, she was waiting for her first flight assignment.

A chill of apprehension crept through me as I drove to the medical center. Soon I was standing at Patty's bedside, gazing down at her face covered in bandages, gauze pads over her eyes. I spoke to her and sensed that she felt my presence. Scott was standing nearby and his voice shook as he talked. A United Airlines pilot with the cool head of someone accustomed to being in charge, he didn't often lose his composure. But on this day he was not in control. Scott asked me to go with him to the hospital's chapel, where we prayed and tried to get our heads around what had happened. Sudden loss is like a punch to the gut, leaving you disoriented, and Scott knew I had experienced it before. I think he felt my ability to get through my recent medical problems and the uncertainty about my future might help him understand why this terrible thing had happened.

After we prayed, I offered some words that I no longer remember but seemed to give him some momentary comfort. We embraced and then I left him at Patty's side, surrounded by her family. Patty died later that afternoon. She was thirty-eight.

After her funeral, I reevaluated my fascination with spaceflight. My mind often returned to the myth of Icarus and his golden wax wings, his errant flight toward the sun, and his fatal plunge to Earth. I have loved Henri Matisse's jazzy rendition of Icarus all my life, including when I was a young child. It reminded me that if you fly too high you will fall. I had learned about the hubris of Icarus and his tragic demise during my sophomore year in high school in Ms. Patterson's Latin class. Years later, on my worst days, I wondered if I had exhibited a similar flaw. Had I been guilty of excessive pride, of too much ambition? If so, would I be able to overcome it? Would I get the chance to do this space thing in my condition, as a severely hearing-impaired astronaut?

I did more soul-searching as spring rolled into summer and I was assigned to the robotics branch in July. I would go to the office and mill around, go to the gym, or just go for a walk. I was recovered enough to appear normal, but it seemed like there were still wires being connected in my head. I tried to understand how to position myself strategically in social events so I could discriminate single conversations amid the din created by a room full of people. The right positioning could enable me to hear a conversation and not appear to be deaf. This was not something I had ever had to worry about while meeting with my astro-buds at Petey's after a T-38 cross-country flight or a training session in the simulator.

I faced another challenge at the launch parties, those gatherings where individuals who had been assigned space flights celebrated with their families and NASA employees. Some of my fellow Penguins were receiving assignments and holding their own launch parties. The "astronaut-only" mission de-briefs didn't help either. We were invited to these sessions to hear and learn from veteran astronauts who shared stories of their missions in space. The de-briefs helped our class bond as a group, but now that wasn't enough for me. I also saw the impact these men and women had had on the public, particularly children and teachers. As I waited and wondered about earning that coveted flight assignment, I saw success occurring for others. Would it ever happen for me?

● ● ●

After my hearing stabilized I started training in the Single System Trainer (SST). The space shuttle electrical systems, general-purpose computer (GPC), and auxiliary power units (APUs) were all systems that I had to relearn after my accident and after supporting the Expedition 1 crew in Russia for two years. It was time for me to get back on the proverbial horse.

I was working through some computer malfunctions one

day in early September when I got summoned to Astronaut Chief Kent Rominger's office. Something had happened in New York involving terrorists, and I could see on his office TV that smoke was coming from one of the twin towers of the World Trade Center. All astronauts were told to leave the premises and sequester themselves in their homes because no one knew if the Johnson Space Center would be the next target. It wasn't out of the realm of possibility, and the presence of the first Israeli astronaut, training at the center with his crew from STS-107, heightened concerns. We were told not to step outside our homes for the next three days.

While we were responding to those instructions, Astronaut Frank Culbertson was high above us. He had just completed routine physical exams of the crew of the International Space Station when the flight surgeon down in Houston shared the grim news. Frank suddenly had the odd distinction of being the sole American not on Earth during the 9/11 terrorist attacks.

In an e-mail to his family and colleagues, he described watching the dark plume rising ominously into the sky from the eastern United States as he was filled with a profound helplessness and isolation.

It's difficult to describe how it feels to be the only American completely off the planet at a time such as this. The feeling that I should be there with all of you, dealing with this, helping in some way, is overwhelming. I know that we are on the threshold (or beyond) of a terrible shift in the history of the world. Many things will never be the same again after September 11, 2001. Not just for the thousands and thousands of people directly affected by these horrendous acts of terrorism, but probably for all of us. We will find ourselves feeling differently about dozens of things, including probably space exploration, unfortunately.

It's horrible to see smoke pouring from wounds in your own country from such a fantastic vantage point. The dichotomy of being on a spacecraft dedicated to improving life on Earth and watching life being destroyed by such willful, terrible acts is jolting to the psyche, no matter who you are. And the knowledge that everything will be different than when we launched by the time we land is a little disconcerting. I have confidence in our country and in our leadership that we will do everything possible to better defend her and our families, and to bring justice for what has been done. I have confidence that the good people at NASA will do everything necessary to continue our mission safely and return us safely at the right time. And I miss all of you very much. I can't be there with you in person, and we have a long way to go to complete our mission, but be certain that my heart is with you, and know you are in my prayers.

Humbly,

Frank

Frank had also just learned that the pilot of the plane that crashed into the Pentagon that day was his classmate from the Naval Academy, Charles Burlingame. It would be another three months before he climbed aboard space shuttle *Endeavor* for the trip home. He had been in space for 129 days.

• • •

Months flew by in a dizzying blur as the country tried to recover and as I continued the struggle to find my place at NASA. I thought that if I continued to work hard and stick to my goals I might become an arm operator on a shuttle mission. But, there were so many steps and approvals to obtain before that could happen.

I continued to train in the SSTs and then started doing fixed and motion-based training, which involved both a static version of the space shuttle and one that moved to give you the sensation that you were launching off the planet. I continued to get my hearing tested at Methodist Hospital downtown and tried to get back in the flow of being an astronaut, albeit one whose future was uncertain. I believe my managers wanted the best for me but didn't know what to do with me. So, they figured, why not let him train? In the robotics branch I had to get proficient on both Canadian robotic arms. I went to Montreal to train to be certified to fly it. I came home and would do qualification runs to support other missions. That meant working in Mission Control Center (MCC), ensuring that if the crew had to perform unexpected robotic maneuvers we would help validate the procedures and I would fly the trajectories to make sure they would operate safely on orbit. This branch position and support role lasted until September 2002, when Astronaut Office Deputy Andy Thomas asked if I would go to NASA headquarters in Washington, DC, to support the new Educator Astronaut Program that NASA Administrator Sean O'Keefe was launching with a new head of education. He knew my parents were educators and that I had a passion for inspiring the next generation of explorers.

I accepted the assignment, but before I moved, Jake came into my life. As a kid, I knew what it was like to have a dog. Back then, though, I didn't have to be completely responsible for the dog's care because I had a family. Caring for King, the family's collie, and Jocque, our French poodle, was a task shared by everyone in the house. Jake would be different, and my sole responsibility.

Jake was a Rhodesian ridgeback and Chow mix, a combination that would explain the dog's athletic build and boundless energy. Jake was an active dog to say the least. Rhodesian ridgebacks could go twenty-four hours without food or water while tracking down lions. It was Jake's energy and unpredictable behavior that prompted his owners to find a new home for the dog. The couple had just had a new baby, another reason to find a new home for Jake. I knew Jake would become my dog. While there were no lions on the loose in my suburban neighborhood, I figured Jake's stamina, intelligence, and loyalty would be great traits to have around the house and on long hikes in the mountains. I brought him home the following week. At that point, Jake was a ninety-pound teenager—warm and affectionate one minute and barking up a storm the next. Initially, Jake had some separation-anxiety issues. Still, we bonded immediately.

The next month, I drove to Washington and moved into the unfinished basement of my godsister, Renee Abbott, an attorney with the United States Patent and Trademark Office. I was going to keep Jake with me, but we realized my travel schedule, with the new job, would make that too difficult to manage. My parents agreed to look after Jake. I had planned to leave Washington during the weekends for occasional visits to Lynchburg, but that did not always happen.

Educator Astronaut

NASA's Educator Astronaut Program was created to reinvigorate the public's interest in the space program by stirring up the passion of students and teachers. I was the comanager of the program with Debbie Brown; she was the educator and I was the astronaut. Fifteen people were detailed to NASA Education headquarters from across the agency to develop the program and kick it off in January 2003. Our team was really close and we worked around the clock from November through December to get everything in place. It was one of the best teams I have ever worked on because everyone was so committed to inspiring kids and teachers to work with NASA. My boss, Adena Loston, was new to the agency and relied on me to help her understand the space part, as well as brief her on the office politics. I traveled around the country meeting educators and students to promote the program. The fact that I was not on flight status was the only sticking point. Some of the kids asked if I had ever flown in space, and when I said no, I lost a little credibility. They said I wasn't an astronaut because "to be an astronaut you have to fly in space and you haven't." The kids didn't know I was medically disqualified to fly, but it hurt because of my uncertainty.

Part of the program's mission was to select between three and six teachers to begin training to become "educator mission specialists," much like Christa McAuliffe sixteen years earlier. Back then, NASA's Teacher-in-Space program generated so much interest among teachers that 12,000 applied for one slot. The prospect of a teacher becoming an astronaut gave the space program a tremendous boost.

McAuliffe was a charismatic choice who would win the hearts of many as the nation's first teacher-turned-astronaut. NASA selected the thirty-seven-year-old social studies instructor from New Hampshire to fly on the space shuttle *Challenger*. Barbara Morgan was selected as her backup and both women underwent training in Houston. But as we all know, McAuliffe never made it to space. On January 28, 1986, a ruptured fuel tank tore the shuttle apart just seventy-three seconds into the flight, killing everyone on board. A shocked nation mourned the loss of the crew, and the space shuttle program came under fire.

"We'll continue our quest in space," President Ronald Reagan said at the time. "There will be more shuttle flights and more shuttle crews and yes, more volunteers, more civilians, more teachers in space. Nothing ends here; our hopes and journeys continue."

Twelve years later my Penguin classmate Barbara Morgan flew aboard space shuttle *Endeavor*. The Educator Astronaut Program was NASA's proof that it was dedicated to sending more teachers to space. By the end of my first year, three more teachers had won slots in the Astronaut Corps.

• • •

I had been working on education programs for NASA for about four months when I decided to make a trip to Lynchburg to see my parents and Jake. It was a Saturday morning when I began the three-and-a-half-hour drive from Washington, DC. I had always loved the way Highway 29 South meandered gently along the Blue Ridge Mountains to the west, and I had been looking forward to the drive. I remember chatting on my cell phone with my friend Rudy King, a NASA colleague, when I noticed Adena Loston was trying to reach me. She was burning up my line, so I switched over to talk to her. The space shuttle *Columbia* was scheduled to land at Kennedy Space Center in Houston, and she was calling to ask why the countdown clock was going up instead of down. Adena was

NASA's associate administrator for education, and she looked to me to explain the intricacies of the space shuttle program from an astronaut's perspective. I suddenly felt sick and immediately pulled off the highway. Countdown clocks count down, not up.

I turned on the radio. It was unclear what had happened. NASA had not made an official announcement, but within a few minutes, news broke that the shuttle had broken up during reentry into the Earth's atmosphere, sending debris in every direction. The unthinkable had happened, again. I turned around and headed back to my office in Washington.

The agency's gears shifted immediately. We needed to take care of our families, and on the night of the disaster I was dispatched to provide support to the parents of *Columbia* Mission Specialist David Brown, a flight surgeon who had been among the crew. David had led the investigation of my hearing loss in the Neutral Buoyancy Laboratory (NBL) two years earlier and I considered him a close friend. As I drove to the home of Paul and Dorothy Brown, a state trooper stopped me at the driveway, sent there to head off the swarm of reporters who had been arriving all day. Just a few hours after the accident, a reporter disguised as a flower deliveryman had surprised Dorothy when he pulled a microphone out of a bouquet and asked for a comment.

Nothing had prepared me for what I was about to go through helping to protect and console my friend's family. I had never met Dottie and Judge Paul Brown, and this was not the way I wanted to see them for the first time. That night, I stayed with the Browns until they went to bed, not able to process the tragedy myself. My friends were gone, and suddenly I wasn't the only astronaut grounded. The entire shuttle program was in jeopardy.

"My son is gone," David's father had said to me. "There's nothing you can do to bring him back, but the biggest tragedy would be if we don't continue to fly in space to carry on their legacy." Hearing his words, I started crying. We all hugged and cried, and I felt such warmth and love for this amazing couple. The judge and his son shared the same sparkling eyes and compassionate temperament. At that moment, I realized why David was such a talented and selfless person. He was always willing to help others, as he did when he doggedly led the investigation to find out what happened to me in the Neutral Buoyancy Laboratory pool, all while training for his flight.

Judge Brown's comments stuck with me and radically changed how I felt about my place in this world and what it means to think of others first. I wanted to do my part to honor his words.

When I got back to the office, I attended an Educator Astronaut Program meeting. I was exhausted and emotionally spent from the sudden loss of my seven friends and the evening I had just shared with Dave Brown's grieving parents. In Houston there were meetings where astronauts were briefed and offered medical and psychiatric care if needed. In addition, they had the benefit of a shared experience to aid in their recovery. As the only astronaut in the education program and perhaps the only current astronaut at NASA headquarters that morning, I felt as if I were on an island. That feeling was heightened when one of my colleague's first words about the accident were "I can't believe this is happening to me." I looked up for a moment, thinking that my bad ear was pointed in her direction and that I didn't hear her correctly. But she repeated it over and over. I was so upset that I wanted to leave the room. She had turned an opportunity to rally the team to honor our fallen heroes into a moment of startling selfishness. We all left that meeting feeling sad and dejected. I felt certain that no one around me could possibly understand what it meant to be an astronaut and have seven of your astronaut buddies die all at once, in a flash.

One of them, Willie McCool, had been very patient with me from my earliest days as a Penguin. Once, he was flying to Florida to do a Space Flight Awareness event, where astronauts give awards to employees and thank them for their contributions to helping us fly safely. Willie asked me if I could go but technically I couldn't. He got me a waiver to go from the chief of the office. We flew the jet down and had a great conversation on the way. While in Florida, he left our rental car in my capable hands while he went to run an errand. I ended up locking the keys in the car. As a result, he was late for an appointment while employees from the rental agency retrieved the keys. During your AsCan year you are barred from doing any type of media. They don't want you to get out there and say something you have no knowledge about, or aren't authorized to say. The media ban forced me to keep my mouth shut at the event while Willie handed out the awards. The next day we drove the rental car to a designated place for them to pick it up and then take us to our T-38. A few hours later, we landed at Ellington Field and for some reason went straight to the ready room where Shelly asked, "Do you guys know where you put the keys to the rental car?" I told her I left them on the visor. "They're looking and they can't find them," she said. I tapped my flight suit and, to my distress, the keys were sitting in my pocket. I had committed all these rookie mistakes but Willie was very gracious and didn't get upset, which was the coolest thing about the whole trip. I really appreciated being his friend and knowing him. His demeanor defined the crew that perished. There was this civility, this kindness that all of them possessed.

• • •

It had been seventeen years almost to the day since the 1986 *Challenger* disaster. The second tragedy involving the space shuttle *Columbia* sent the NASA community into indescribable grief and changed the course of the nation's space program.

A two-year investigation revealed that a large piece of foam dislodged from an external fuel tank during takeoff and damaged the wing of the shuttle. Nothing seemed amiss until the shuttle re-entered the Earth's atmosphere on February 1, 2003, when the damaged wing caused the shuttle to break apart. As the nation grieved its second shuttle disaster, Congress began questioning whether space travel was too risky and if the shuttle program was the best way to continue human space flights. A commission was set up to examine the root cause of the *Columbia* disaster, and the finger pointing on Capitol Hill and within NASA began in earnest. The president had to decide if NASA would continue the shuttle program or scrap it because of its old technology and rising concerns about safety.

In truth, nobody really knew what would happen next. The five shuttles—*Atlantis, Challenger, Columbia, Discovery,* and *Endeavor*—had flown 111 successful missions. But the

Challenger and the *Columbia* disasters would become the two flights seared into the public's memory. Though we all knew the shuttle program would begin winding down eventually, this was not the ending we wanted.

The abrupt end to the shuttle program coincided with my own uncertainty. All at once, I was dealing with the loss of friends, the realization I might always be hearing impaired, and the painful probability that I would never fly in space. A few days later, President George W. Bush went to Johnson Space Center to deliver a memorial speech in honor of the *Columbia* crew.

"Our whole nation was blessed to have such men and women serving in our space program," he said. "Their loss is deeply felt, especially in this place, where so many of you called them friends. The people of NASA are being tested once again. In your grief, you are responding as your friends would have wished—with focus, professionalism, and unbroken faith in the mission of this agency."

The president went on to proclaim that the space program would go on. "This cause of exploration and discovery is not an option we choose; it is a desire written in the human heart," he told the gathered mourners. "We are that part of creation which seeks to understand all creation. We find the best among us, send them forth into unmapped darkness, and pray they will return. They go in peace for all mankind, and all mankind is in their debt." His words resonated with me. They gave me hope that the program would endure.

As a crew astronaut casualty officer, whose responsibility is to support colleagues' families in times of crisis, I attended the memorial services of the *Columbia* astronauts with NASA leaders and other astronauts, crisscrossing the country on the

NASA plane. The entire agency was in shock, not only from the catastrophic loss of so many of its own, but also because of the possible shutdown of future shuttle missions and the uncertainty of the nation's space program.

What I didn't know was the toll it was taking on me. From the minute I'd learned of the *Columbia* accident I'd been helping the families of the crewmembers. I was there to help them endure unbearable grief, but I didn't allow myself to mourn at any of the memorials because that wouldn't help anyone. I was in serious caregiver mode.

Several weeks passed before I had flown to Houston, the first time since the shuttle accident. By that point, I had attended memorial services for most of the *Columbia* crew. I took a commercial flight from Washington, DC, to the airport in Houston and rented a car. I will never forget the feeling I had as I turned onto NASA Parkway and headed toward the Johnson Space Center entrance. There on the side of the road, in front of the space center sign, was a spontaneous memorial of hundreds and hundreds of American flags, Bibles, bouquets of flowers, letters, and handwritten prayers.

I pulled my car to the side of the road and the grief that I had long avoided caught up with me. I cried uncontrollably, for my deceased buddies, their families and friends, the space program's future, and for myself—a broken, grounded astronaut.

It didn't register at the time, but it turned out the agency's chief flight surgeon, Dr. Richard Williams, had been watching me closely as we flew from town to town to see the families, assessing how the flying was affecting my ears. I noticed he routinely took notes on takeoff and landing, but I didn't know what for.

A few weeks after we returned from traveling to one memorial after another, Williams summoned me to his office at NASA headquarters. I had no idea why he wanted to see me, but when I walked into his office, he stood and extended his hand to me.

"Leland," he said. "I'm going to sign a waiver so you can fly to space."

8

Training for Space

Normally, astronauts know when they're going to have a significant conversation about their status because it takes place according to a timeline. For example, new classes report over the summer so that they can start training in the fall. After going through the entire selection process, applicants know they will receive a thumbs-up or thumbs-down call in May or June. I had no clues to prepare me for my conversation with Rich Williams. Becoming medically qualified to fly was the last thing I thought we would ever discuss.

Williams gave me the news, but I suspect that Jon Clark and John Locke had also played a pivotal role in his decision. They had always wanted to push the envelope in terms of my recovery, advocating experimental procedures that might have helped me heal faster. My conversation with Williams thrilled me; I was finally going to get a chance. I thought of Jeannette's prophecy and marveled that it was now coming to pass. I called my dear friend Mary and told her the news. I think she was even more excited than I was because she saw it as an affirmation of faith and testimony. Diligence, prayer, and belief in oneself could indeed make good things happen.

Shortly thereafter, I spoke with Bill Ready, astronaut in charge of human exploration. I told him I had done all I could at NASA headquarters, especially since the Educator Astronaut Program candidates would be transitioning to the selection process in Houston. In late May, I left HQ in my Jeep Grand Cherokee, pulling a U-Haul trailer behind me. I picked up Jake in Lynchburg and continued south to my home in El Lago.

In mid-June, Jake and I loaded up and hit the road again. The trip included visits to the Grand Canyon, Hoover Dam, Yosemite National Park, and other places of interest in the Southwest. I borrowed a storage unit to go on top of the Jeep from Duane Ross, head of astronaut selection for thirty years and the man who saved me from falling off my chair during my interview. Behind the wheel, I was pumped by a new sense of optimism. My waiver didn't clear me to get back in the T-38, or resume training in the buoyance laboratory pool, which meant I was ineligible for long-duration flight and the required spacewalks. Still, I was back in line for a coveted space shuttle flight, and that alone was reason enough for joy.

As I drove, my thoughts drifted to the past six months I'd spent in the Educator Astronaut Program and the loss of the *Columbia* before returning to my new opportunity. I could still hear Judge Brown's remarks about the importance of continuing the space program despite our catastrophic losses. He was right, of course. The best way to honor our fallen comrades was to take their place in the cosmos and embody, as they did, peace, love, and hope for a better world.

Jake and I took in the beauty of the natural environment as we visited the sites on my list. Along the way, connections we made with nature and humanity confirmed my hopeful sentiments. And we had adventures in the midst of it. In Yosemite, we met two women who had just graduated from college and were driving across the country. That evening, we sat by a campfire and reflected on the future while looking up at the night sky seeing the twins Castor and Pollux and Orion's Rigel and Betelgeuse. Jake lay by my side and slept as we engaged in pleasant conversation. Later, after I had crawled into my tent and fallen asleep, Jake woke me. He was not barking but trying to open the tent with his nose by putting it on the zipper. He wanted to get outside and protect me from a bear that had wandered into camp. I kept him in the tent, and in the morning we assessed the damage. Park rangers told us the bear not only had gone after improperly stored food but also had attempted to break into cars in search of things to eat. I was glad that Jake had shown himself capable of knowing when to bark and when to stay silent. He may be a city dog but I guess he got his instinct from his genes.

Recharged by my experiences in the great outdoors, I returned to Houston, and from September to December I served on the Astronaut Selection Committee. I was now on the other side of the table, listening to why applicants thought they had the right stuff to join the Astronaut Corps. The committee usually consisted of about twelve members, a mix of astronauts and senior NASA leaders at Johnson Space Center. I remember when one candidate came in and I initially thought we would have to coax his accomplishments out of him. Instead he unleashed them on us, relying heavily on the use of "me" and "I" as if teamwork had played no part in his successes. He told us his wife had advised him to check his humility at the door. It was pretty bad advice.

It was such an honor to make decisions that influenced the future of the U.S. space program. I couldn't believe how accomplished some of the candidates were, with multiple PhDs, advanced pilot and scuba qualifications, and some that had hiked ten of the world's ninety-six mountain peaks with an elevation of at least 14,000 feet. We met Type A personalities, ready to do whatever it took to climb into a rocket and take off to the heavens. Others had communication skills that didn't match their other achievements. I remember many times locking eyes with other board members and wanting to ask particular candidates, "Why the heck would you say that to us if you are trying to get hired?" I wondered if I sounded that way when I interviewed. Judging by the way things turned out, I guess not.

In January, my NASA duties included serving as an Astronaut Potted Plant (APP) during a trip to Washington to help promote President Bush's vision for space. I use the term to describe when astronauts are required to lend their presence to speeches that refer to the program. In such instances, the president or NASA official would gesture in our direction when he or she came to the relevant passage. Speakers and audiences like having astronauts in the room. The allure of space heightened considerably when actual space explorers were standing nearby in the flesh. Fortunately, I shared the stage with a diverse pool of astronauts, including Ed Lu (Asian), Stephanie Wilson (African American), Ellen Ochoa (Latina), John Grunsfeld (Caucasian), and Peggy Whitson, a Caucasian woman who would soon become the first female commander of the space station. President Bush outlined a course for additional missions, including the development of a new manned space vehicle and a return to the moon. "With the experience and knowledge gained on the moon," he said, "we will then be ready to take the next steps of space exploration: human missions to Mars and to worlds beyond."

My next trip was to a conference in Orlando that NASA co-sponsored. The National Association of Minority Engineering Program Administrators (NAMEPA) focused on developing strategies to increase diversity in the engineering industry. In my presentation, "Living Your Dreams," I showed images of young African American kids morphing into astronauts. The stories I shared included an account of Charlie Bolden's journey to space and the top ranks of the space program.

Charlie grew up in Columbia, South Carolina, and was a gifted student who had hopes of entering the U.S. Naval Academy. His only problem was he needed to be nominated by a member of Congress, and then senator Strom Thurmond, a staunch opponent of federal efforts to bring racial integration to the South, refused to do it. "No way are you going to get an appointment from me to go to the Naval Academy," he said.

Eventually, Representative William Dawson, an African American congressman from Chicago, nominated Bolden, who went on to earn a BS in electrical science. He later became a Marine Corps officer and a naval aviator who flew more than one hundred combat missions during the Vietnam War before getting a master's degree in systems management.

In 1980, he applied to become an astronaut at the urging of Ron McNair, his friend and mentor who died in the *Challenger* explosion, just two weeks after Bolden returned from his first shuttle mission aboard the *Columbia*. Four shuttle missions later, Bolden left NASA and returned to active duty with the Marine Corps. He retired in 2003, but a few years later, he returned to lead NASA as its administrator. Ironically, one of his first duties was the thankless task of overseeing the retirement of the space shuttle program.

I hoped sharing these stories about achievers like Bolden would inspire kids to see themselves in places that hadn't always been open to us, like the space program.

In June, I traveled a lot farther. I was headed to Tokyo, Japan's capital and largest city. The greater Tokyo metropolitan area consists of more than 37 million people, making it the world's largest metropolitan area. Fortunately, I was scheduled to go to Tsukuba and the Japan Aerospace Exploration Agency to become certified to operate a new robotic arm scientists there had developed. The flight into Tokyo was easy enough. So was the hour-plus bus ride from Tokyo to Tsukuba.

The training was essential if I ever hoped to get into space and work the shuttle craft's robotic arm to move payloads— the cargo, equipment, or spare parts for the space station— between the spacecraft and the station. What I learned from meeting the scientists, engineers, and space officials during the training sessions was that there are key similarities that unite us, no matter where we're from or our cultural differences. Everyone wants the best for their children, food on the table, and a roof over their head. They desire dignity and respect. Most just want to be heard.

I did manage to get a couple of days off for a trip back to

Tokyo, where I met my friend Koichi Wakata and his family for sushi. Wakata is a Japanese engineer and astronaut who participated in four NASA space shuttle missions, a Russian *Soyuz* mission, and a long-duration stay on the International Space Station. The small restaurant where we had lunch stood out for the way it served the food. The dishes came out on a little conveyor belt, and you grabbed what you wanted. The unagi and yellow tail were pretty good. So was the green tea sponge cake.

I had my big 35-millimeter SLR camera and several big lenses with me, and after lunch spent time walking around the city. Eventually I wound up near the residence of Japan's prime minister, where his procession was about to enter. Although the police were holding visitors at a distance, enthusiastic citizens motioned for me to come closer. Seeing my big camera and lens, the security assumed I was part of the foreign media, and they let me pass the barricade to get closer to the prime minister's entourage.

Wakata got a good laugh out of that. I always called him "The Man" because everywhere we went people who recognized him would bow down to him as if he were royalty. At the time, the two of us were learning Russian and would say, "You're so *den'gi*," which is Russian for money. The line was from the movie *Swingers*. It was a little surreal to be in Japan telling each other "I'm so money" in Russian.

I spent much of 2004 on the road, but some of my travel was devoted to pleasure, not business. The year before, while traveling with Jake, I had spotted a unique-looking vehicle at Stinson Beach near Muir Woods in northern California. It was a van with four-wheel drive and a pop-top. I looked inside and saw a fully decked-out camper, but not like the shaggin' wagons of the 1970s with the dice and lime-green shaggy carpet. It was more reminiscent of my dad's bread truck. On my trip back, I got on the phone with my good friend Karen and we learned more about Sportsmobile, the company that had made the van. With the company's help, we started designing my own customized camper. I picked it up in July 2004, and Jake and I set out to explore once again.

Our country includes an abundance of marvels and splendors, many of which are too distant or inaccessible for most Americans to witness firsthand. In some respects they compared to the wonders of the cosmos, and I was quite grateful to be able to explore them. My appetite for wandering had continued to percolate ever since my father instilled it in me long ago, and driving around the country with Jake reinforced my certainty that such tendencies were part of the national personality. During restful moments, I read from John Steinbeck's *Travels with Charley*. Writing about his campsite neighbors, he noted, "I saw in their eyes something I was to see over and over in every part of the nation—a burning desire to go, to move, to get under way, anyplace, away from any Here. They spoke quietly of how they wanted to go someday, to move about, free and unanchored, not toward something but away from something. I saw this look and heard this yearning everywhere in every state I visited. Nearly every American hungers to move." More than forty years after Steinbeck's trip, I could see that some of his observations still rang true.

Among the places we visited that summer was Columbia Point in Colorado. The U.S. Department of the Interior had dedicated a 14,000-foot mountain to the fallen crew of the *Columbia* and renamed it after them with a commemorative plaque. We also spent time in southern Idaho, where I had lunch at a diner and was served by a waitress who had never met a black person. I enjoyed my visit, but I took care to stay away from the northwest portion of the state, which was known by the locals for its white supremacist group activities.

My work for the space program picked up again in the fall of 2004. In October, I flew to Brussels for a conference put on by Astronaut Frank De Winne, who would command the International Space Station on my last shuttle flight. Frank was very passionate about education and had been given resources to start a program in Belgium about space education. The topic of the conference was Space Serving Education. I had made an impression with the work done with the Educator Astronaut Program, so I had been invited to share the results of our efforts. I even got the chance to speak before the Belgian Senate, sharing my experiences serving as an astronaut educator.

By January 2005, my fellow astronaut John Herrington and I were in Alaska. We met with teachers and education officials in a few remote villages. Ours was a goodwill endeavor under the auspices of the Educator Astronaut Program (EAP) and another program called Explorer Schools to promote interaction with students and teachers. Much has been discussed about the immense distances astronauts travel in space, but we often cover a surprising amount of ground while carrying out less glamorous but important missions on Earth. And those bring their own rewards, such as getting to meet people all over the world, including children who may be so inspired that they will someday follow in our footsteps.

• • •

By summer, instead of palling around the great American Southwest with Jake, I began kayak training conducted by the National Outdoor Leadership School (NOLS) at its site in Palmer, Alaska, with fellow astronauts Jim Halsell, TJ Creamer, Paolo Nespoli, Terry Virts, and Robert Thirsk. Our instructors were Andy and Chris with NOLS Pro. After packing our gear and going through orientation, we traveled by boat to Prince William Sound, where we were dropped off on a small tidal island near Nassau Fjord. We learned to choose a campsite and practice Leave No Trace (LNT) tactics to maintain it. Leave No Trace is a national partnership between federal land management agencies, educators, conservation groups, and private businesses to preserve the land for future generations. The tactics are important to future astronauts who must be diligent in keeping their environment inside a spacecraft as sterile as possible. The habits we learned from Leave No Trace can make the difference between life and death in space.

The next day we learned how to enter a kayak and quickly get out of an overturned one. Kayaking long distances daily with our team challenged us all physically and was used to test our resolve under pressure. It helps trainers evaluate the behavior of astronaut candidates in harsh conditions, particularly as we may have to spend extended time in outer space. We initially did four and a half miles and camped on Nassau Fjord. We started each day with a team briefing and discussion of what we wanted from the course. In addition, we learned important maxims such as, "If the map does not agree with the ground, the map is wrong."

My journal from the trip includes observations about the weather and an itemized list of knowledge I wanted to gain from the experience:

1. Mastery or at least a strong comfort level in the solo flotilla. Understand pace/fatigue factor and carrying capacity. For dog, 2-person exploring, what would be the better choice, kayak or canoe? Types, etc.
2. Wilderness medicine
3. Ropes & knots
4. Navigation: celestial

However, my favorite entry simply reads, "Taking extravagant pleasure in being alive."

• • •

Later that month, I was reminded of the fragility of life and the speed with which joy can give way to desperation when Hurricane Katrina devastated most of New Orleans and huge portions of the Gulf Coast. Displaced by the storm, many people hurriedly stuffed their belongings into their cars and left the area for safer communities, including Houston. Others stayed to face the full fury of the storm. The sadness and uprooting caused by the storm's destruction and the broken levees drove home the limitations of progress in the modern world. On the one hand, we possessed the technical savvy to create vehicles and satellites capable of reaching distant regions of the heavens. On the other, we still haven't figured out the solutions to some very basic questions on Earth, such as how to keep every one of God's children, safe, fed, and warm.

Witnessing the genuine difficulties that other people experience helped distract me from my personal, far less troubling questions involving whether I would ever get to space. Getting that clearance waiver was one thing; getting assigned to a mission would be quite another. I continued to work at NASA and, as the calendar pages flew, I found myself contemplating an eventual return to Virginia.

I thought about how nice it would be to go back there someday and start to build a little homestead. I had talked to my dad about me getting a few acres in the country to build on. That dream turned into an idea about raising kids there and educating other kids, perhaps operating a camp in which they would learn skills that incorporate art with science, technology, engineering, and mathematics. Art is an important part of STEAM, the educational concept to encourage youngsters to appreciate scientific and technical instruction.

One day my dad called me and told me he had bumped into a man named Mr. Wheeler and had asked him if he knew of anyone that had any land for sale. The man replied that he did and my dad went with him to his house. The funny thing was, when my dad walked into his house, Mr. Wheeler's wife realized she had gone to school with my father. The possibility for the transaction had been set.

Before Thanksgiving of 2005, I flew to Lynchburg and walked the fields in Appomattox, Virginia, where the Civil War ended, and close to the land we hoped to buy. The visit to Appomattox made me realize that soldiers had probably camped out on that very farm. After my dad and I walked the land, we both liked what we saw and made an offer. I came back in July 2006 and closed the deal. It was so peaceful there. I could see huge clouds meandering in the sky, casting shadows on the serene fields. I quickly thought of Serenity and named it that, soon after closing.

I was the proud owner of ninety acres of farmland, half wooded and the other half soybean fields that at one time were used for tobacco. Streams, fox, deer, and bear tracks were peppered all over the property. There was an old tobacco barn and a log cabin, now planked over, that was the house that Mr. Wheeler was born in. He had permitted the Bushwhackers Hunting Club to use the land, and when I purchased it, I let them cross my property to hunt. However, I determined that such activities would have to stop once I began to build, and certainly could not continue with children present. One Christmas, the club gave me tenderloin from the freshly killed deer as a token of gratitude.

By the time I took over the land it was nearly overgrown and starting to go to seed. Mr. Marston was a farmer who lived a few miles down the road in Red House, Virginia. He cleared the land, planted soybeans, and kept the yard clear down by the old house. I had no experience with farms, but my mother grew up on one. It was only a farm because of the fields; I had no livestock to worry about. My dad would drive down every so often to walk the fields and check on everything. It gave him a sense of pride. I lived in Houston and would only get to Lynchburg on holidays, but I would find peace every time I would visit Serenity, our family farm.

The idea behind Serenity fit my demeanor. Extreme emotions of elation or despair weren't part of my makeup, despite my life's unforeseen twists and turns. I had low points, of course, but often resolved them by turning to faith lessons I had been brought up to believe in. At times, I often consulted 2 Timothy 1:7, "For God has not given us a spirit of fear and timidity, but of power, love, and self-discipline." Perhaps more than any other factor, living a life of faith had prepared me for that moment in June, when Kent Rominger, chief of the Astronaut Office, called to tell me I was going to space.

The news was unexpected. It came totally out of the blue. I was in my bedroom when I got the call. Jake was sleeping on the floor.

Kent told me I was assigned to STS-122 on the space shuttle *Atlantis*. Our primary mission was to transport the Columbus Laboratory and attach it to the International Space Station. It would be the first permanent European research facility in space. Stephen Frick would command the shuttle, and I would join a crew consisting of my fellow mission specialists Stan Love and Rex J. Walheim, along with Hans Schlegel and Léopold Eyharts, from the European Space Agency. Alan Poindexter was our pilot. I would have the responsibility of installing the lab once we reached the space station.

I remember jumping up in the air and letting out a loud scream, pumping my fists in the air, and saying, "Yes!" Jake woke up and wondered what was going on, and I got down on the floor and hugged him and we started playing. I called my folks, who were very excited. Word spread through my hometown. My sister, Cathy, threw a party at the Lynchburg Public Library and the mayor gave me the key to the city. The Big Blue Crew showed up, as did my good friend Butch Jones. My old coaches, Mark Storm, Jimmy Green, and Rufus Knight, were there, too.

• • •

The projected launch date for STS-122 was December 2007. Our mission was to deliver the European Space Agency's $2 billion Columbus Laboratory to the International Space Station. The space agency regarded the Columbus as its future center of activities in space and had been waiting ten years to have the twenty-three-by-fifteen-foot research laboratory installed.

It would be my job to connect the modules. I recall walking into a meeting in September 2006 with about a dozen or so members of the European Space Agency and hearing my mission flight director introduce me. "This is Mission Specialist Leland Melvin. He is going to install the Columbus," he said, and the room burst into applause. I felt pressure knowing that all of them had been waiting so long for this moment, and I was their guy to do the job. One of the project leaders, a European flight controller, turned to me as I walked out of the room, and in a thick German accent, said, "Don't screw it up."

Training requirements for a mission are laid out almost a year in advance, according to a precise template. The instructors all meet, check off everyone's tasks, and indicate when they think the crew is ready. The commander and the chief of the office talk and reach an agreement. If a crew's vehicle isn't delayed for some reason, they are slated to go when the shuttle is ready.

For my first flight, most of the training took place in Houston, where the fixed- and motion-based space shuttle simulators are located. The facilities also include a virtual reality lab, which allowed us to integrate robotics operations and enabled extravehicular activity astronauts to don virtual reality goggles, gloves, and portable life support systems before immersing their bodies in the virtual worlds of the space shuttle and the space station.

In the Neutral Buoyancy Laboratory we practiced robotics and extravehicular activity procedures to properly reach, feel, and move objects. The goal was to make sure what we did here under supervision could be duplicated in space. There was a submerged space shuttle, space station, and both robotic arms, which gave us the opportunity to practice ground-control approach maneuvers close to the station structure. If done wrong, these maneuvers can kill an astronaut. They can be crushed by the arm if wedged between it and the structure, or the arm can actually break the astronaut's glass visor.

It was important for Rex, Stan, or Hans, the spacewalkers, to give me clear calls regarding the direction and distance they wanted to travel while attached to the robotic arm. I got to a point in the training where I knew where they needed me to be to perform the task and put them in the right place to get their jobs done. They had to trust that I was looking out for all their body parts through the cameras and would actually call them if I could not see sufficient clearance. If anything looked unsafe we would say, "Stop motion," and reassess to ensure safety. I loved flying the robotic arm, especially when I was in the zone and could put all of my training and skill to work, employing the rotational and translational hand controllers to make the arm move and dance freely in space. As lead for robotics, that's how I briefed all participating crewmembers before every practice run. And it paid off in training and on the flight.

Other training took place offsite. In October, I joined Steve, Alan, Rex, Hans, and Stan for a NOLS survival skills course in Canyonlands, Utah. We carried eighty-five-pound backpacks for ten days, traversing washed-out roads and withstanding the treacherous effects of a one-hundred-year flood. Our NOLS instructors pushed us to our limits to see how we reacted as a team in adverse conditions. We all had our breaking points. My training with three-a-day football sessions had prepared me for those all-day survival-type sessions. The military guys had undergone similar training, especially the special forces types. We wanted to be ready for whatever the space shuttle, space station, or space itself threw at us. We couldn't allow personality conflicts or discomforts to get in the way of the mission: Deliver the payloads and get home safely.

In the spring of 2007, I did more training, including a stint at the Memorial Hermann–Texas Medical Center in Houston to learn the basics of emergency medicine. Every space shuttle mission has a designated medical officer. Often it just so happens that one of the astronauts on a mission is a medical doctor, and that was the case with my second mission, STS-129. On that flight, there was Dr. Robert Satcher, a Harvard-trained orthopedic surgeon with a doctorate in chemical engineering from MIT. We had no such person on my first shuttle flight, so I volunteered to take on the task.

Given the limitations imposed by my incident in the Neural Buoyance Laboratory five years earlier, I figured I should make myself useful in every way possible. I wasn't squeamish and liked the science of the human body. As a professional athlete I had developed an appreciation for physiology and anatomy—I knew what the body could withstand.

The medical officer is responsible for handling any medical problems and emergencies that come up during the mission, along with routine stuff like doling out the sleeping pills that some astronauts take to help them get through the zero-gravity nights. The training took eight weeks and covered everything from inserting a catheter (training I unfortunately had to put to use on the mission) to stitching up wounds and giving injections.

The most incredible part was that I was assigned to work

with Dr. Red Duke. He had been a young surgical resident at Dallas's Parkland Memorial Hospital on November 22, 1963, when President John F. Kennedy was shot by Lee Harvey Oswald as the presidential motorcade rolled past the Texas School Book Depository. He was the first doctor on hand to receive the president before he had to turn his attention to saving the life of another patient brought in that afternoon. Though he didn't realize it immediately, that other patient was John Connally, the Texas governor who also took a bullet that day but survived.

Dr. Duke was a force. He was a skilled physician in the operating room who possessed a folksy, country-doctor bedside manner. He liked to hum Willie Nelson songs around the ER; the two apparently were close friends. "He was a world-class surgeon trapped in a Texan's body," Congressman Ted Poe once said, describing him as "John Wayne in scrubs." Luckily for NASA, he revered the space program.

On my first day training in the ER, a teenage boy was wheeled in with a large gash on his leg, the result of a car accident. "Go stitch him up, Leland," Dr. Duke barked. The cut was several inches long, but I first had to stitch the muscle underneath. If there's one thing they train you in the Astronaut Corps, it's how to act confident in the face of uncertainty. "How many of these have you done?" I remember the boy asking, his big eyes staring into mine. "Enough," I responded in a casual voice. The truth was that I had, if you consider cadavers in that meaning of "enough." It was part of my training.

Under Dr. Duke, I learned how to find arteries, draw blood, and put clips on blood vessels to keep patients from bleeding out. We started on cadavers, but as time went on and our confidence grew, we worked on live patients in the ER. NASA had a way of putting astronaut trainees in stressful situations, and I couldn't think of too many more stressful than working on a Friday night in a big-city hospital emergency room.

My medical training also included learning how to insert and use catheters, a skill that can save lives in outer space. On Earth, gravity exerts force on a full bladder to let you know that you have to use the bathroom. There is no gravity in space, where after a few days a full bladder can explode.

On that same day that the car-crash patient asked about my experience, I had to insert a tube into a man's stomach, through his nose. That was a first for me, too. But, perhaps the most dramatic trial by fire that day came in the form of a drunken young woman covered in tattoos and piercings. She had been the driver in a collision that had killed her friend. We had to insert an IV but her veins had collapsed, and a rattled young resident was having no success inserting the needle into her jugular vein, the only accessible blood vessel. "Leland, take over," Dr. Duke said. I knew I just had to do it, and my success earned me a reputation in the ER.

Not every day of training was so dramatic. Back at the space center, the crew sometimes celebrated a good run in the simulator by going to local watering holes like Boondoggles or the Outpost to let off some steam. Mostly, though, we trained, and training continued until the day the mission began. An engine-cutoff sensor problem pushed our planned December 2007 launch until February 2008. We were able to rest for a short while but mostly we took "refreshers," exercises that kept our skills sharp and up to date.

About a week before heading to Florida, we sequestered ourselves in Johnson Space Center crew quarters for a week. Any visitors had to be cleared by a doctor because no one wanted to take any germs into space. Our families weren't with us but our robotics simulators were, and we practiced until we flew to Cape Canaveral for the final five days before launch.

Launch day: I remember all of us in the elevator leaving the second floor of astronaut crew quarters, starting to walk out, seeing the lights and cameras, our friends shouting and screaming. My high school chemistry teacher Cornealea Campbell and her son Cornel were among those on hand to cheer us on. I remember hearing him yell my name as we prepared to enter the silver Astrovan to ride to the launchpad. Typically, an astronaut could be assigned to a mission after basic shuttle and station training, which takes about a year and a half. Because of my medical situation, my path was far from typical: I ended up training for nearly ten years, starting from my AsCan days in 1998.

In the van, I looked over at Rex, whom I would be sitting next to in the shuttle. We exchanged grins. Finally, it was about to begin.

9

The Final Frontier

I am about to take the ride of my life. Our seven-member crew is strapped into the shuttle *Atlantis,* according to seat order. Dex and Steve went first, then Rex and me on the flight deck, followed by Stan, Leo, and Hans on the mid-deck. Suit techs and astronaut support personnel make sure we don't bump anything while getting in our seats. They connect the crewmembers to the oxygen and the chiller that will keep us cool in our suits over the next three hours. The techs hand us our checklists, notepads, and pens that are tethered to us, enabling us to pull them back if we drop them. Actually everything is tethered, in case our gloved hands make us clumsy. After getting situated, we perform a suit check to make sure we have no leaks.

The clock at Cape Canaveral counts down to the moment

of liftoff, when the shuttle's three main engines roar to life. Seconds after the main engines start, the solid rocket boosters will ignite and lift us through the heavy parts of Earth's atmosphere. In less than three minutes, the boosters will be jettisoned from the shuttle after their fuel is spent. Still on the ground, we have a ten-minute window to fix any problems that may be detected in the shuttle's hydraulics, electrical, or fuel valve systems and still launch.

If the launch is forced to occur during that window, it's important to make sure there's enough fuel, proper weight, and perfect weather conditions to chase the International Space Station. It seems I have been chasing goals placed before me—the NFL, a career in engineering, and now a space station orbiting high above the Earth.

I fist-bump Dex, Steve, and Rex as the safety systems are armed, the balance and pressure systems checked, and the water activated. The thrust of the main engines tilts us forward and back, a motion we NASA folks call "the twang." Suddenly there is the wondrous thunder of the solid rocket boosters roaring to life.

Imagine you're in a sports car going about one hundred miles per hour. Our acceleration was one thousand times more intense. We were pinned in our seats, feeling three times our weight on our chests. I remember laboring to breathe during the two minutes before the solid rocket boosters were finally jettisoned. At that point, I thought, *OK, we're heading to space*.

A minute later, we reached 10,000 miles per hour, rocketing over the east coast of the United States with the Atlantic Ocean shimmering in the background. Another five minutes passed as we reached 17,500 miles per hour, fast enough to shut off the engines and jettison the external tank.

When the main engines cut off, I felt I was no longer chasing space. I had arrived. I had made it to that spacious habitation, the phrase that came to mind when I first looked down from space and saw our home, the Earth. A reporter asked me after the *Atlantis* mission what was it like to be up there. I initially spoke about floating and seeing things that weren't attached to anything in the shuttle floating around us. The talk quickly turned to that magnificent view of Earth. I saw the planet for the first time without borders. I thought about all the places on Earth where there's unrest and war, and here we were flying above all that, working together as one team to help advance our civilization. That was an incredible, incredible moment for me.

From space, our views of the rivers, seas, and oceans—with all their various shades and pigments—would challenge any painter's vocabulary. There aren't fifty shades of blue to describe the Caribbean because its blues from space are so intense. Turquoise, indigo, azure—I quickly ran out of words that would do it justice. I didn't think about color a lot before going to space. Although I appreciate sunrise, sunset, the blackness of the night sky, and the blinding white of snow-capped mountains, I hadn't anticipated such a stunning array of blue water.

But we weren't up there to take in the sights, amazing as they were. Hans and I unstrapped ourselves from our seats and prepared to take images of the shuttle. I shot video and he shot stills. In one image we captured the fan-shaped spread of light that is the propellant-residue flares from the external tank. The tank itself plunged toward the Indian Ocean. After we reached orbit, we set about converting the shuttle from a rocket ship into our home away from home. First we had to open the payload bay doors before activating the shuttle's robotic arm, the device I had been training so long to operate. On flight day two, our first full day in orbit, we used the arm to grapple an inspection boom. Moving it up the side of the payload bay, we scanned the orbiter to make sure our heat shields hadn't been damaged during takeoff. We also started prepping for the first of our three spacewalks, and the chance to put all that extravehicular activity training to good use.

On flight day three, we connected with the International Space Station. We began the day about forty miles behind the space station, checking off our rendezvous protocol and using laptops, a laser ranger, and other tools as we maneuvered to get below the station. By the time we completed our procedures and approached docking, the shuttle and the space station were moving slowly toward each other, while each was moving in orbit around the Earth at 17,500 miles per hour.

The station is essentially a series of interlocking modules in low Earth orbit. After initial assembly of the station—which began in 1998, the year I became a Penguin—most modules and compartments, usually research laboratories of some sort, have been delivered and installed via shuttle missions like ours. Once in operation, the Columbus lab, which we were delivering, would be used for experiments in biology, biomedical research, and fluid physics.

After contact and capture, we joined Steve, our commander, in briefly celebrating our safe rendezvous. A NASA image of the moment shows me congratulating him with a camera in my hand.

We opened the hatches and entered Node 2, where we greeted space station commander Peggy Whitson and her crew, Yuri Malenchenko and Dan Tani. The month before we arrived, Peggy and Dan had participated in a seven-hour, ten-minute spacewalk, replacing a motor at the base of the one of the station's solar wings. Dan would be heading home with us on our return flight and Leo would be replacing him on the station. We were pleased to be able to stretch our legs, even while adjusting to our new environment.

Most of the modules on the station have four sides and they're put together in a way that enables the crew to work continuously on flat planes, either a wall, a floor, another wall, or the ceiling. All they have to do is turn and their frame of reference changes. Because objects in orbit are in a continuous state of freefall, handrails, tethers, Velcro, and other devices are employed to keep objects securely attached to the work surfaces.

In addition to all the workspaces, there is a kitchen and two bathrooms. Gene Roddenberry, the creator of *Star Trek*, used to joke that there were no bathrooms on the *Enterprise* because the crew just set their phasers on disintegrate to get rid of waste. Bodily functions are handled differently on the space station. The facilities are quite small, containing a tiny toilet for a number two and a corrugated hose topped with a yellow funnel for number one. Going to the bathroom for the first time was an adventure because (1) we didn't train for it much on the ground, and (2) in zero gravity, *everything* floats. Fortunately, both devices offer a bit of suction to keep things going in the right direction. A little curtain provides some privacy, but not much.

On the station, sleeping quarters are similarly confined. Each sleep station is like a tiny phone booth/office, with sleeping bags to the walls, a computer, and room for a few books. On the shuttle we would strap our sleeping bag with bungees and float in our bags in a place that a few hours before bedtime had served as the kitchen.

There are so many unforgettable aspects of life in space, including the experiments, the robotics, and the spacewalks, but I think my most memorable moment took place when Peggy and her crew invited us to have dinner over in the service module. "You guys bring the vegetables, we'll bring the meat," they said, and we all congregated around the small table, with some floating above and others below. There we were, French, German, Russian, Asian American, African American, listening to Sade's silky vocals and having a meal in space. Out the window we could see Afghanistan, Iraq, and other troubled spots. Two hundred and forty miles above those strife-torn places, we sat in peace with people we once counted among our nation's enemies, bound by a common commitment to explore space for the benefit of all humanity. It was one of the most inspiring moments of my life.

While the space smorgasbord with Peggy's crew included Russian and international cuisine along with canned beef and barley, most of our meals consisted solely of typical American fare. Many people associate astronaut food with that freeze-dried ice cream sold in museum gift shops. In truth, you won't find that on the space station, but you will find thermally stabilized and irradiated food that tastes much like the entrees served on Earth. Some food needed only to be heated while others required the addition of water. My favorites included beef brisket, mac and cheese, and string beans with almonds. M&M's and Raisinettes made great snacks, with the added bonus of being fun to play with. We trapped the tiny treats in water bubbles and gobbled them up as they floated by.

Staying properly nourished and fit was critical to our success at performing the jobs we had to do. The effects of zero gravity on the body made self-maintenance part of our daily routine. Without the pressure of Earth's gravity, the body begins to do funky things. For example, every vertebra gets room to move, stretching the spine. I'm five foot eleven on Earth but I was six feet on the station. After my spine elongated, when I went to bed on the first night I felt some pains in my lower back. I had to curl up in an attempt to alleviate the discomfort. The heart also changes in space. Its gets smaller and changes shape because it doesn't have to pump as hard to pull the blood up from your feet. Without gravity, our bones change shape, lose calcium, and become more brittle. As a preventive measure, we worked out on a treadmill specially designed to help us combat loss of bone density. (We also had an exercise bike and a resistive exercise device or weightlifting machine.) Some astronauts experience intracranial pressure changes that push on their eyeballs, changing the shape of the eyeballs and forcing the astronauts to wear glasses in space. We kept different prescriptions of glasses on board just in case someone's vision changed.

Meanwhile, our major objective was to install the Columbus Laboratory. It began with a spacewalk. Before passing through an airlock and a hatch to enter the vacuum of space, Rex and Stan put on suits with bulky backpacks that on Earth would weigh about 300 pounds. Their equipment included oxygen, heating, cooling, carbon dioxide removal systems, and a computer. Once in space, Stan attached a grapple fixture that enabled me to grab and move Columbus with the fifty-eight-foot robotic arm. Working with Dan and Leo at the robotics workstation, I could look at monitors to maneuver the big shiny module. It was very slow work that required a lot of configuration to line it up just right. We had beautiful views of the Earth through the aft window as we pulled Columbus into its berth in Node 2. Following the installation, we had outfitting tasks to do, including removing launch locks and installing handrails. Our second and third spacewalks involved more work with the Columbus, including attaching external payloads. I operated the arm for those procedures and others, such as switching out a nitrogen tank on the space station and retrieving a broken gyroscope for repair and reuse.

When we weren't busy moving and installing payloads for the space station, we'd take in the light shows. For example, when we were over the Earth's southern hemisphere we'd see this green glow of particles hitting the atmosphere. The colors were different over the northern hemisphere—purple, yellow, and blue. I had been told about the cosmic rays that would pass through the vehicle and hit my optic nerves, making me think I was seeing flashes of light, even though that was not really happening. The flashes were like sunbursts of different colors popping in my eyes and in my head.

Sleeping brought a different kind of light show. It was a pretty incredible experience because we didn't have the sensation of lying down; we floated inside our sleeping bags as we dozed. We slept amid a whole din of pumps and motors, so many whirring around us; the noise made it seem like we were in a factory. On top of all that, my dreams were so vivid from the stimulus of the day. Behind my closed lids, I saw blues, greens, whites, whether they came from the ocean, whether they came from the sun, or whether they came from flashes in my head from these high-energy particles. The colors intertwined with my dream state and I sometimes saw alien forms and green clusters of light moving and dancing in a way that made me think of little green men on Mars. At one point in the mission, a huge, inside-out cheeseburger, dripping with grease, began to float through my dreams. I was just chomping on this burger, a juicy contrast to the irradiated food I had been consuming in real life.

There was no birthday cake on the shuttle either, but that didn't stop me from celebrating my own big day. It happened when a surprise party organized by the Astronaut Family Support Office brought my parents, sister, and a host of friends to a conference room at NASA Langley. Through a video hookup I saw all their beaming faces, surrounded by blue balloons and noisemakers, as they gathered around a cake. My parents wore gray sweatshirts with "Atlantis" emblazoned across the front. I had just finished a long, challenging day operating the arm, and seeing their faces made me happier than they could have realized.

"A lot of prayers are going up to you," my dad told me.

"A lot of prayers are coming down to you," I replied.

The "party" ended with the whole group serenading me. Rudy King, a coworker at NASA Langley who often played basketball with me, blew out the candle.

"This is really special, guys," I said. "If I cried, the tears would just float away. I'll save those tears for when I get home."

Five days later, our mission was completed. It was time to return to Earth. In the history of human spaceflight, there had only been a little more than five hundred people who had been given an opportunity to go off-planet, and I had been one of them.

We said our goodbyes to Peggy, Leo, and the station crew

as Dan prepared to join us for the flight back. We did one last 360-degree fly-around of the station to take more photographic documentation before beginning a de-orbit burn, a small adjustment to our orbit that sent us skimming into the planet's atmosphere. We started to bleed off the enormous speed we'd attained when the rocket boosters had sent us hurtling from the pad at liftoff. Our displays indicated a speed of Mach 25, twenty-five times the speed of sound. We could look out the overhead windows above us and see a 3,000-degrees-Fahrenheit hot pulsing plasma that was only about three feet from us as we entered the atmosphere. Fifty minutes later, twin sonic booms announced our imminent arrival to the families and friends awaiting us at Kennedy Space Center.

Our drag chute deployed and jettisoned our forward landing gear; we touched down on runway 15. We took off our pressure suits and, after completing our last few procedures, carefully disembarked from the shuttle. I took a last look around. I had spent ten years chasing space. And I had finally caught it.

The best part of coming home was seeing my family and friends and the immense joy and pride I felt while walking on terra firma again, not floating but feeling the Earth beneath me. It was a beautiful moment. The next morning, waking up in my own home, I had moments of disorientation in which I wondered, *Where am I? Am I in space? Am I on the ground? How am I going to move? How am I going to eat? Will I have to throw food in the air and fly and get it?* To walk on the beach was the most incredible thing. When you look at the horizon it helps regauge your gyros and your inner ear. It helps you know exactly what's up and down and what's right and left. Driving again required some adjustment because I'd been traveling at 17,500 miles per hour, an experience that changed how I thought about speed. I would get in my car and think, *Hmm, I'm going sixty miles per hour, oh, I'm going one hundred miles per hour, oh, that's nothing.* I'd always been enthusiastic about food, but it tasted even better after my time in space. It was especially gratifying to come home to a wonderful meal and share it with loved ones without having to chase it when it bounced off of something in zero gravity. I could just pick it up with my fork and relax in the certainty that it would arrive straight to my mouth.

Going to space is transformative on so many different levels. I tell future explorers that when you work with others, make sure you work together as a team. Learn to see all people as potential space travelers together no matter what language they speak, no matter what they look like, no matter what food they eat, and know that we're all in this together. Work hard and share the fun.

When you look at the Earth from the vantage point of space, our planet looks like a little blue marble. Seeing our world from that vantage point cognitively changes you. My orbital shift happened after breaking bread with my space station crewmates and my shuttle crewmates. It showed me how close we are as countries, as races, as a species. I marveled that on Earth we have all these distances and separations and geographic boundaries, but they vanish quickly in the weightless interior of the space station.

I'd always been low-key, relying on the strength of my upbringing and my faith. I could easily find solace and reassurance by walking the restful fields of my farm. Yet, I was even calmer when I returned from space. I realized that the things that I thought were a big deal were no longer so important.

The routine, day-to-day frustrations, like someone cutting

you off in traffic, paled in comparison to the experience I'd just had. Space also inspired me to think about how we reach that next generation, how we prepare others to take our place and go off somewhere else and do much better things. It was a notion I would return to with increasing regularity in the years to come.

• • •

Each member of a shuttle crew is allowed to carry mementos into space. We packed some into our personal preference kits, one for each crewmember. Others were stored inside the Official Flight Kit, a two-cubic-foot container. My personal items included two NFL jerseys (the Lions and the Cowboys, of course), a beloved Curious George book from my childhood, jazz bassist Christian McBride's *Live at Tonic* album, and some songs I had composed myself in my little home studio. Among my most meaningful mementos was a work of art created by youngsters at Project Row Houses (PRH). Based in the Third Ward, one of the oldest African American neighborhoods in Houston, PRH aims to use art to transform the social environment. In the spring of 2007, Melissa Noble, a friend and longtime fixture in Houston's cultural scene, introduced me to Rick Lowe, the project's award-winning cofounder. His vision of community renewal resonated with me, and I subsequently visited the group's Arts/Education Program. After I spoke with the children about my upcoming shuttle mission, PRH worked with Houston artist Jenna Jacobs to help nearly fifty neighborhood youngsters, all between the ages of five and fourteen, create a three-by-five-foot quilt reflecting what space meant to them. They worked all that summer to complete it before my flight, and I was able to take it with me when the mission launched in February. In April 2008, nearly two months after my return from space, I went back to PRH to talk to them about my trip. As usual, I felt a special connection with my youthful audience as I recalled bearing witness to the wonders of the cosmos. Perhaps best of all, I was able to present them with their quilt. It had traveled nearly 5.3 million miles and circled the Earth 203 times.

In June, I joined the SST-122 crew on a much shorter journey than our previous expedition. We headed to Germany to celebrate our successful installation of the European Columbus Laboratory and Leo Eyharts's historic accomplishment as the first European astronaut to live on the space station for an extended stay. We also paid tribute to Hans Schlegel, the German astronaut whose spacewalk with Rex had secured Columbus to the space station. We arrived in Berlin and visited historic sites such as Checkpoint Charlie, the name western Allies gave to the Berlin Wall crossing point during the Cold War. Our diverse international crew provided a contemporary contrast to the historic enmity such landmarks recalled. I remember our crew taking a picture with Angela Merkel, the German chancellor. She had invited us to meet with her at the Federal Chancellery in Berlin to brief her on the technical and scientific results of our mission. Such briefings often require translating difficult research concepts into language more readily understood by the masses. Chancellor Merkel, a trained physicist, required no such translation.

A month later I was in Israel to participate in a pro-am tennis tournament. My appearance was part of an event promoting the Israel Tennis Center's (ITC) first annual Israel Open in memory of Ilan Ramon, who perished in the *Columbia* disaster. The first Israeli astronaut, Ilan had been my classmate and friend. The ITC is the largest children's tennis program in the world and helps children develop socially, psychologically, and physically in a multicultural environment. The ITC's fourteen centers are located primarily in disadvantaged neighborhoods where they serve a variety of youngsters, from those with special needs to gifted athletes. Among the most talented was Anna Smashnova, the professional with whom I was partnered in the tournament. Born to a Jewish family in the Soviet Union, Anna won that country's youth championship in 1989, when she was just fourteen. A year later, her family moved to Israel at the invitation of Freddie Krivine, one of the founders of the ITC.

I did more than play tennis while I was there. Accompanied by Ilan's widow, Rona, I met with Prime Minister Ehud Olmert. On another day I went to Jaffa to visit the "Sky is the Limit" ITC coexistence project, a program designed to promote bonds between the various ethnic communities in Israel. I also attended a gala to benefit the center, with many of the country's best professional tennis players on hand. I went to Masada, one of Israel's most popular tourist attractions, and even floated in the Dead Sea, which I had seen and photographed from space.

Because of my experience with the Educator Astronaut Program, I was also called on to speak to groups of children. It gave me great pleasure to share my stories of space travel at the Sderot and Ashkelon Tennis Centers, where I met with Ethiopian children and others who lived in cities that had been repeatedly hit by rocket fire from Gaza.

"No matter what happens in our lives, we have to keep moving forward," I told them. "We have to keep doing our best, no matter what the circumstances. It is about your heart, dedication, and spirit." Wherever I went, I told my young audiences that few people I knew possessed as much of those qualities as Ilan, Israel's fallen astronaut. People readily embraced me because of my ties to him. In 2003, the Israeli government named seven hills in the Ramon Crater in honor of Ilan and the six U.S. astronauts who died on the *Columbia*. Five years later, the memory of his heroism was as strong as ever.

• • •

Back in Houston by August, I resumed training as a Cape Crusader. It sounds like Batman's job in Gotham City but it's actually a position created to support the Astronaut Office at Cape Canaveral during missions. The tasks are related to operations, and include helping as part of the close-out crew, getting astronauts in their respective seats in the shuttle, and doing communication checks with Mission Control Command. Also, in the event of a terrorist strike, Cape Crusaders may need to help the SWAT team. As a result, we trained in firearm shooting, developed familiarity with flying a helicopter, and otherwise prepared to step in and save the day where required. Tracy Caldwell and Paolo Nespoli trained with me, along with Barry "Butch" Wilmore, with whom I would soon be working closely.

By soon, I mean right away. Kent Rominger, the chief of the Astronaut Office at the time, called me into his office and told me I'd been assigned to STS-129, scheduled for liftoff in November 2009. Butch was assigned to pilot the shuttle for what would be his first mission. We'd be joining Commander Charlie "Scorch" Hobaugh and mission specialists Randy Bresnik, Mike Foreman, and Bobby Satcher. After Kent gave me the surprising news, I went looking for the other crewmembers to congratulate them. Mike and I had been classmates so we had a great relationship. I had been on the selection board for Bobby and Randy, and it was cool to fly with people I had a hand in choosing. I was excited and shocked to be named to the mission since it had not been that long since I had flown.

I have been asked if being part of the Astronaut Corps affected my sense of time. It certainly does in the sense that you know you only have so many weeks to train to be perfect. If you are not perfect you can kill yourself, someone else, or the entire team and also end the space program. Having flown before in no way diminished my sense of high stakes during the months of training preceding my second mission.

I got a welcome break at the end of the year when I went home to Virginia for Christmas. Spending time with dear friends and relatives recharged my emotional and spiritual batteries. The atmosphere was festive in Halifax, where my mother was born, as my aunts and uncles broke bread and told stories of days gone by. The farm, which had been in the family for many years (and still is), was at the center of many of their tales. Later that month my dad and I walked the land, a sojourn that was becoming a holiday tradition for our family.

The new year started with considerable promise. My friend Melissa Noble, whose friendly overtures led to me taking a quilt to space, introduced me to the artist Elaine Duigenan. We met at Indika, one of my favorite Indian restaurants in Houston, where I learned about some photographs she was working on. The Micro Mundi series captures images of the trails that snails make when they are eating algae. When photographed from above, these tiny images look like aerial photos of Earth. I was so enchanted with her project that I took one of her pieces on the *Atlantis* in November and photographed it floating in the weightlessness of space.

The high point of the month was my trip to Washington, DC, to attend President Barack Obama's inauguration. Watching alongside wives of the some of the crewmembers who rode on the NASA float in the inaugural parade, I saw a black man take the nation's highest oath of office while his countrymen cheered. Like so many others, I was filled with the most incredible sense of pride. I remember hearing Itzhak Perlman and Yo-Yo Ma lead a quartet performance of the song "Air and Simple Gifts." Their sensitive performance moved me to tears even though it was ten degrees outside. Few things could reasonably be expected to match the thrill factor of such an extraordinary and historic occasion, except maybe going to space.

• • •

My training adhered to a typical routine over the next several months, full of exercises, repetition, and critical attention to detail. One welcome respite took place in May, and my fellow astronaut Suni Williams was very much a part of it. Our love of dogs was one of the many things Suni and I had in common. A few years after I got Jake, a second dog that looked almost exactly like him showed up on my lawn. I had watched Scout wandering around the neighborhood day after day. He had no collar, he was nearly blind, and I quickly learned he had heartworms. He must have had a home somewhere nearby but his owners clearly didn't care about him. So there I was with two nearly identical Rhodesian ridgeback mixes who came to me by accident, separately. While Scout was relatively mild-mannered, Jake was considerably more excitable. He protected me like a grizzly bear defending her cub, a particular challenge when people came to the door. He sometimes tried to bite people who came too close.

Meanwhile, Suni's Jack Russell terrier, Gorby (named after former Soviet leader Mikhail Gorbachev), was also having discipline problems. He had the unfortunate habit of going after dogs bigger than him. So far he had escaped any real consequences, but Suni was frustrated.

The NASA Public Relations Department caught wind of our dog problems and decided to turn it into publicity. And on a sunny day in May, Cesar Millan, star of National Geographic Channel's *Dog Whisperer*, drove down to Houston in a fancy RV that he parked in front of my house in El Lago. That's when we filmed "Cesar, We Have a Problem."

The voice-over began as I opened my front door to Cesar. Jake and Scout went nuts and jumped on him. Both dogs tried to slip past my legs and into the front yard. The camera panned to Suni standing in the hallway.

"Astronauts Suni Williams and Leland Melvin have been flawless in space but are experiencing technical difficulties with their dogs here on planet Earth," the unseen narrator said. "Cesar likes to tackle the toughest problem first so he begins with Leland and Jake." At one point, I thought Jake was going to attack Cesar. What our guest didn't know was that Jake would often run to the window and bark at the garbagemen. The men, seeing my dog was stuck in the house, responded by beating on the window and harassing him. The men were Hispanic, and my fear was that Jake would associate his tormentors with Cesar, especially when Jake began snarling and showing the Dog Whisperer his teeth.

When Jake started to act up, Cesar saw my alarm and told our viewers, "That joyful, calm, passionate astronaut went out the window." Cesar's advice: Be calm, just like I had been in space. If Jake sensed a fearful attitude in his owner, he'd react accordingly. If he sensed a calm unruffled attitude in me, he would take the appropriate cue and settle down too. Cesar's patient counsel proved helpful. In time, Jake settled down. Cesar went on to give more advice that helped Suni and me better handle our dogs. The episode aired the following year.

A summer hiatus took the form of a big reunion with family and friends at Holiday Lake State Park in Appomattox, Virginia, just down the road from the courthouse where Robert E. Lee surrendered to Ulysses S. Grant. I'm sure neither man could ever have expected we would send humans to space one day, and most likely could not even begin to fathom a man with slave ancestry, one that owned property in Appomattox no less, eventually flying in space. My parents, cousins, and friends from past and present joined me in walking the fields of Serenity after the picnic. As we walked, I felt grateful to be surrounded by and immersed in family and community. To love and be loved was its own kind of protection, armor for my upcoming journey. I had only four months before heading off to the heavens again.

September brought a collaboration that would produce another work of art to take with me in space. Instead of a beautiful patchwork quilt or inventive photograph, the result this time was a musical recording. When I began to write poetry in the aftermath of Hurricane Katrina, I never imagined that my efforts would lead me to Pharrell Williams's studio. I met the Grammy Award-winning singer-producer at the fortieth anniversary of spaceflight at the Udvar-Hazy Center in our home state. Our discussion sparked his imagination, and led to a trip to Miami, where he had me read my poem, "Exploration." It begins,

The journey of a city made in haste
Many people frown about the waste
Cluttered minds and empty souls
Wonder how we will one day behold
The world without treasures found today
In the atmosphere far away
Floating around the heavens we see
Advancing the future with harmony
Seen in galaxies miles away
Solutions to the crisis in the world today

Listening and working intently, he crafted music to the lyrics. The finished product arrived in the mail a week later.

I did one final hike in Muir Woods in late October, after completing training in the simulator at NASA's Ames Research Center in Moffett, California. I had walked the same terrain after the *Columbia* accident to calm my soul following the loss of my friends. This time it was to gather my wits as I prepared to leave the planet again. I hiked from the visitor center, ten miles round-trip, along the Dipsea Trail down to the beach where I had seen the Sportsmobile six years before. Then I had been in a totally different state of mind. Now I was about to head back to the cosmos. I devoted the hours to exploring the outdoors, communing with nature, and finding peace within. I ate my favorite meal at the Parkside Café, the house salad with ahi tuna, delicious. It had been a tradition to eat it on my three visits to Stinson Beach. Each journey I had taken since receiving my second assignment had served a purpose, whether to connect with planet Earth, to embrace friends and family, or even to behold a new president. My experiences in the Astronaut Corps had taught me that you never know what will happen from day to day. It's best to experience life as fully as possible while you have the time, health, and opportunity.

• • •

"Liftoff of space shuttle *Atlantis*," launch commentator George Diller announced, "on a mission to build, resupply, and to do research on the International Space Station." We launched at 14:28 EDT on Monday, November 16. We shimmied in our seats, feeling the familiar twang as the rockets fired. And we were off.

Diller's job description was accurate. We had a variety of tasks on our to-do list, most of them focusing on staging spare components outside the space station. You may have noticed by now that NASA is fond of acronyms and abbreviations. I mention this because our primary payload was the ExPRESS (Expedite the Processing of Experiments to the Space Station) Logistics Carrier ELC-1 and the ELC-2. The carriers contained a good deal of hardware, including gyroscopes, nitrogen tank assemblies, pump modules, and a high-pressure gas tank— some 30,000 pounds of replacement parts. Installation of the equipment would require three spacewalks and frequent operation of the robotic arm.

A few hours after we got to space, we were all working hard. A couple of the rookies were feeling sick and moving slowly but I felt fine. Suddenly I began to make strange sounds, like a turkey gobbling. Before I knew it I was barfing my brains out, even though I had never felt sick. I aimed for the emesis bag designed for such occasions but didn't quite hit the target. The green fluid bounced and hit the University of Richmond hat I was wearing before it migrated to my eyebrows and mustache. I peered through the veil of vomit that had been trapped by my hat and could see Butch and Scorch looking down at me from the flight deck, their faces frozen in astonishment. Like a football player who has been knocked to the sidelines, I just wanted to shake off my troubles and return to the field. Welcome back to space.

• • •

We docked on day three and met with the station crew. Commander Frank De Winne welcomed us, along with Jeffrey Williams, Robert Thirsk, Nicole Stott, and cosmonauts Roman Romanenko and Maksim Surayev.

Later that day, Randy and I used the shuttle's robotic arm to lift ELC-1 out of the payload bay. We handed it over to the station's arm operated by Butch and Jeffrey, who then permanently installed the carrier to the outside of the station. The following day Bobby and Mike performed our first spacewalk and completed their tasks without a hitch. That night, however, a false depressurization alarm sounded and woke us, but flight control teams on the ground determined there was no danger to the station or crew. It was the first of three alarms that would sound during our mission, all of which proved false. Had they signaled real emergencies, we would have been prepared to handle them.

Everything about our training drives home the point that space is an incredibly dangerous place. If you do something wrong it's game over. To keep everyone safe, we routinely practice emergency procedures. The biggest emergency is depressurization. If something's coming at you and hits the space station, it's going to cause a breach. Even something as small as two centimeters in diameter can do a lot of damage. If that first alarm had been genuine, everyone would have gone to their respective vehicles and closed the hatches. We would have had only so much time to get in that vehicle, secure it, and be ready to go home. Another situation we train for is fire. If a fire alarm were to go off, our computer screens would display the module where fire is suspected. Then we would go to the location to see if we could identify smoke or flames or anything coming out.

Solar flares present another kind of danger. They occur when the sun kind of burps every so often, sending out dangerous high-energy particles. If there's a solar flare all the astronauts float up into a space at the top of the station (we call it the doghouse). It has a ton of water bags surrounding it and the water will absorb the radiation and keep it from coming into our bodies. Then we just wait until Houston calls us and says everything is clear. As astronauts we're trained not to be alarmed by things. We're trained to ask, "How will I fix this? How will I do all the steps required to make sure that we come home safely?"

Surprises can still happen despite all our practice, but even then our training kicks in and enables us to respond accordingly. Experience can be equally helpful.

On Extra Vehicular Activity 3, for instance, Butch Wilmore and I were supporting Bobby and Randy with the robotic arm. We were moving an oxygen tank from ELC-2 that needed to be installed outside the airlock. Butch and I got a little behind in our ops and while we were moving the arm in for the grapple, we had not configured the business end of the arm. It contained three wires that would wrap around the target grapple pin on the payload and allow us to attach it to the arm. Butch was flying and I was supporting and started to set up the grapple. It appeared that we were not ready for capture, but I told Butch to keep moving the arm and it was going to be OK. My confidence had been bolstered from my experience successfully working the arm during my first mission. That was a great moment between the two of us because we saved precious time by capturing the tank without having to stop and reset everything.

Extra Vehicular Activity 2 had been eventful too, but for a different reason. When Bobby and I worked together to move ELC-2 out of the shuttle payload bay and hand it off to the space station's robotic arm, we helped resolve a long-standing historical injustice. Forty-three years after racism forced Ed Dwight from the astronaut program, we became the first African American men to fly together on a shuttle mission.

For a historic moment our "first" was pretty subdued, although Bobby and I had made up shirts that said PTAP (Power To All People). We thought back to the 1960s and 1970s when activists chanted "power to the people" in protest against racism and discrimination. Looking down at our planet, we were moved to raise the bar and ensure that all people were empowered. Seeing the world without geographic boundaries really puts things in perspective and makes one wonder why there is so much division, hatred, and malice.

We shared these sentiments the next day on *The Tom Joyner Morning Show*, a popular syndicated radio program with a huge African American audience. Bobby and I fielded questions from the host while reaching out to young listeners.

We talked about some of the experiments we were conducting with researchers from Texas Southern and Delaware State, two historically black colleges. And we offered a message especially for younger listeners. "They can achieve anything they put their minds to if they believe in themselves and stay determined," I told Tom. "Anybody can do this job. You just have to be focused and determined and just make it happen."

Every crew has their own chemistry and we started to get ours early, before our rendezvous with the space station. Scorch told a joke about two friends going waterskiing on the boat called the *FishBro'*. They were first-time skiers and really didn't know what they were doing. When the one guy asked the other if he was ready, his friend replied, "Make it happen, Captain." The boat accelerated, but the slack in the rope launched the rookie skier into the air. The joke was kind of stupid but from that point on before we did anything we would always start with "Make it happen, Captain." For example, while operating the arm, when I asked Bobby if he was ready to move to a specific location to start service on an end effector, he would say, "Make it happen, Captain."

We had gone to space with six crewmembers but we were coming back with seven. Nicole Stott had been in space for ninety-one days and we were her ride home. It was so funny how the chemistry changed when she came on board the shuttle. It went from a locker room with sailors to a civilized, dignified place. She raised our shuttle aesthetic to a whole new standard, motivating us, as Mike said, "to mind our p's and q's."

Our to-do list left us room for less intense moments, not least because we were able to complete several tasks ahead of schedule. As with other missions, we all brought carefully selected mementos to space. Randy's items included a scarf worn by the famous American aviation pioneer Amelia Earhart. Albert Louis Bresnik, Randy's grandfather, had been Earhart's personal photographer from 1932 until July 2, 1937, when she tragically disappeared. My own artifacts included Elaine Duigenan's Micro Mundi photograph and a T-shirt from Pharrell William's BBC Ice Cream clothing line. I also brought along a recording of "Exploration," the song we'd composed together. Bobby recorded me floating Pharrell's rocket logo while playing the song in space.

My future plans continued to take shape in my mind as we completed our mission. The rhythms of "Exploration" lingered like a refrain in my head: "Floating around the heavens we see / Advancing the future with harmony."

While showcasing Elaine and Pharrell's art in such a scientific setting, I had seen my own notion of advancing the future evolve. My interest in developing a Science, Technology, Engineering, and Mathematics foundation to educate and inspire people had expanded to include art, turning STEM to STEAM. I could do a lot more with it once I returned to Earth.

10

Educator

Following our landing at Kennedy Space Center, the ladies at crew quarters prepared a Thanksgiving feast with turkey, stuffing, cranberry sauce, and all the trimmings. Breaking bread with our families, we had much to celebrate. We had successfully installed two Express Logistics Carriers (ELCs—unpressurized attached payload platforms for the space station), and we brought Nicole Stott back from the station. I slept really well that night, unlike my return from my first mission. That time, staying in crew quarters alone, I awoke in the night and thought I was still in space. As with my first flight, I spent the following morning walking on the beach. Making figure eights in the sand and looking at the horizon helped me get oriented and distinguish between up and down. My sister, Cathy, and my friend Mary joined me on my stroll. We bumped into Mike, my fellow crewmember, and his wife, Lori, and later we saw Bobby Satcher walking along the shore. Nearby, Nicole's sister played with her dogs in the waves. Seeing them frolic in the foamy surf made me miss my own canine companions, Jake and Scout.

Later that morning the crew boarded a jet and headed back to Ellington Field, where we received a heroes' welcome. Years later at an astronaut reunion, I would recall that enthusiastic greeting while listening to Michael Collins of *Apollo 11* fame. In his keynote speech, he told us that none of us were heroes. If your head gets big and you parade around as if you have a big "H" on your chest, he said, you should remember that you were just doing your job.

The next month our entire crew assembled at a Houston Texans football game. We brought home the jersey of the team's leading receiver, Andre Johnson. I had caught a few balls in space while wearing it, after which the whole crew signed it. However, returning the jersey to its rightful owner wasn't the highlight of the day, at least not for me. The best moment occurred when we stood next to former president George W. Bush on the sidelines in our blue flight jackets, singing "The Star-Spangled Banner." During my football days, I'd always felt patriotic every time I heard the anthem before going on the gridiron to do battle. But there in Reliant Stadium, while I stood with my crewmates and the athletes arrayed nearby, the song's familiar refrains offered a stirring coda to my athletic and astronaut careers.

I began 2010 at NASA HQ in Washington, where I'd been asked to support the Summer of Innovation campaign. The program was designed to help underserved and underrepresented students overcome the "summer slide," in which they were likely to lose ground academically and have to play catch-up when the next school year began. I saw it as a chance to use my blue astronaut suit to inspire the next generation of explorers, especially those who didn't always have opportunities in the STEM fields. It was going to be a one-year detail, but the heat and humidity of Houston had convinced me that the East Coast would be my ultimate home. I bought a brownstone on Capitol Hill and went to work.

Between tasks at HQ, I went to Maui in March to receive an award of excellence at the NFL Players Association convention, where the NFLPA was really trying to focus on life after football. I was one of a group of former players lauded for their work beyond the playing field. NFL great Emmitt Smith and other retired players were on hand, along with Muneer Moore and other former Richmond Spiders who had made it to the league. They responded to my presentation about my career as a scientist and astronaut with a standing ovation. As a scientist, I know that predicting outcomes is based as much on probability as on cause and effect. It isn't always possible to isolate the one factor that led to a particular result. Still, as I stepped forward to receive my award, I couldn't help rethinking, however briefly, my journey backward through time, from an NFL event in Maui, through two missions on the space station, to a football field at Heritage High. It had all begun with a dropped pass.

In March, I also met with Thomas Kellner, a German-born artist who had attended the launch of STS-129. With the assistance of my good friend Laura Rochon, I had helped Thomas gain access to Mission Control. He was beginning work on a book that he wanted to call *Houston, We Have a Problem*, inspired by Mission Control's effort to bring the *Apollo 13* astronauts home. I convinced Thomas to change the title to *Houston, We've Had a Problem*, and he eventually published the book. My collaborations with Pharrell, Elaine Duigenan, and Thomas reinforced my realization that art formed the creative thread that tied civilizations together. In providing the common cultural language that unites humanity, it is as valuable and necessary as science, technology, engineering, and mathematics.

By May, the Summer of Innovation program was getting under way. I learned that the associate administrator for education had been let go. If I expressed interest in filling the vacancy, I'd be contending for my former boss's job, which might look questionable to some of my colleagues. In addition, candidates with doctorates in education were usually hired for the job. Despite my reservations, my friend Charles Scales, NASA's associate deputy administrator, encouraged me to apply. In June I called my former space station commander Peggy Whitson, who had become the chief of the Astronaut Office. I asked her if I would have another chance to fly. She thought about it for a few minutes and said no. I could not fly long-duration missions on station because of my ear condition, and we had only a few shuttle flights remaining. Peggy told me she could not offer me anything as good as a senior executive job with NASA. When talk turned to the open associate administrator's position, Peggy was emphatic.

"Go for it," she advised.

I turned in my application and headed off to NASA's Jet Propulsion Laboratory to kick off the Summer of Innovation program with Charlie Bolden, who had become head of the agency in 2009. I recruited some folks from the Corps to help us, including Stephanie Wilson, along with Mike Foreman, Bobby Satcher, Butch Wilmore, and Randy Bresnick, most of whom had been in space with me in November.

Football players weren't the only athletes I was privileged to associate with in 2010. That summer, Venus Williams invited me to the National Press Club in Washington, DC. She was there to sign copies of her new book, *Come to Win: Business Leaders, Artists, Doctors, and Other Visionaries on How Sports Can Help You Top Your Profession.* I had been honored to contribute a chapter alongside such luminaries as Earvin "Magic" Johnson, Soledad O'Brien, Vera Wang, Bill Clinton, and Condoleezza Rice. Venus had asked me to write about "the power of visualization." The topic fit conveniently with my own athletic development. My dad had always coached me to think about the football route, the tennis stroke, the correct form when running. He reminded me to go through the motions in my mind, away from the field or court. Dal Shealy, my football coach at Richmond, had reinforced that technique.

Venus was truly a class act in person. She is no less impressive on the court, where her style of play demonstrates the value of grit and sheer will in pursuing championships. Like many of her admirers, I've always loved watching her and her sister, Serena, crush their opponents. They began on a court in Compton, California, not the likeliest place for tennis champions to launch their trajectories. Thinking about their humble roots took me back to my childhood, when I competed in tennis matches against Linkhorne, the powerhouse middle school where my parents both taught. My father beamed with pride when I beat Linkhorne's top seed, who happened to be the son of my parents' doctor. His play had been honed with private lessons at the country club while my own was shaped on the public courts. Like Venus's dad, Richard Williams, my parents knew the value of grit. They stressed its importance not just on the field but in the classroom, too.

I saw the Summer of Innovation as a way to encourage grit in kids who may have demonstrated it outside of school but hadn't necessarily shown it in class. In August, I was able to implement a nontraditional partnership with NASA Education when we got the opportunity to do a science hologram with the musician and actor Mos Def (now Yasiin Bey). A company that does powerful optical visualization, Obscura Digital, was able to film Mos teaching the science of sound waves and display holographic images of the two of us talking on stage. Our dialogue went something like this:

LM: It's amazing when I think of the parallels that exist between music and science, especially when it comes to sound waves. Mos, can you share some of the science behind the music?

MD: One of the things we know about sound is that it travels by waves. Let's take a look at a picture of a sound wave to see how its shape affects what we hear. In music, long wavelengths create low-pitched sounds and short wavelengths make high-pitched sounds.

I went on to tell our audience about sound waves in space. Mos Def followed with a timely reminder:

MD: There's science all around us. Without advances in science and technology, we wouldn't be able to enjoy radio, television, the Internet, or YouTube, or some of the common things some of us take for granted.

LM: The jobs you will have in the next twenty years haven't even been imagined yet. You will be using technology that hasn't even been invented. We need you to become the inventors. You all have the qualities needed for innovation: curiosity, creativity, vision, courage, and enthusiasm. What you need is the math and science skills to back up these qualities that you already have.

We simultaneously showed viewings of our new tool to a diverse group of students and parents at the Chabot Science Center in Oakland, California, while I was in Florida displaying it to kids and parents at the Tom Joyner Family Labor Day weekend. With special effects and the added sparkle of celebrity, it was a revolutionary way to engage our next generation of explorers.

In October, after enduring a rigorous application and interview process, I became NASA associate administrator for education. I was excited but missed the connection with Houston and the familiar rituals of training exercises and flights aboard jets. My new post was more of a bureaucratic job. It would require long hours testifying on the Hill, scrubbing budgets for cuts mandated by the federal Office of Management and Budget, and remaining responsive to the desires of an administration full of competing missions.

My coming-out party was a conference where we invited leading thinkers in education, philanthropy, outreach, theater, and publishing to see how we could make a difference in education. We had a great lineup of speakers, including Mae

Jemison, Nichelle Nichols, and Charlie Bolden. After setting things in motion with inspiring talks, we rolled up our sleeves to see how we could all work together to improve learning outcomes for our nation's young people. Another one of our speakers was author and "why" guru Simon Sinek. He spoke on the importance of starting a project by asking why, followed by how and what. His talk resonated with many of us, and we eventually called on him to help NASA determine its "why."

I became cochair of the White House Committee on STEM Education (CoSTEM) in November. The committee was tasked to develop a five-year strategic plan for the federal government to share common practices and better ways to leverage its $2 billion investment in STEM. It was a difficult undertaking because many thought the exercise was simply a smokescreen to consolidate agencies' efforts and ease the way for budget cutting. I got a break from all the negotiation and diplomacy in April 2011, when I went to a conference at the headquarters of Lego in Billund, Denmark, where I had the pleasure of meeting Stephan Turnipseed, an engineer turned educator and Lego evangelist. His remarks amounted to a sermon about the importance of motivating kids to build and create with their hands.

NASA, partnered with Lego, developed educational content that involved astronauts assembling Lego Mindstorm Robotics kits in space to help kids see the differences between their device on the ground and how it would behave in microgravity. I remember we did an exercise at the conference where we had to assemble a machine that had to move across the floor to a designated position, but we had to do it without using the wheels in the kit. It was a way to demonstrate creative problem solving. After the demonstration, the conference leaders showed us how eight-year-olds were able to do the same task in a quarter of the time. The youngsters' relative speed illustrated our tendency to purge the creativity out of our kids by making them memorize passages, facts, and figures in preparation for standardized testing. The older children get, the less opportunity they have to be creative, unless they are in an environment that fosters that type of skill set. Hands-on, project-based learning is something that both Stephan and I believed in. Stephan had a farm also, and we talked of our shared desire to motivate and inspire young leaders through the beauty of the great outdoors.

By that point in my career I'd given more speeches than I could count. I'd spoken in various cities, states, and countries, before every kind of audience. You might think that it gets old but it really doesn't. Sharing my ideas with new groups continues to be exciting, and carries the added bonus of stimulating my own thinking. In June I spoke at the United Nations in Vienna, Austria, where I discussed our efforts to bring nations together through space exploration and education. One night I joined a panel of astronauts from Russia, Japan, Malaysia, Switzerland, Germany, Spain, Italy, Canada, and China. The gathering reminded me of my first dinner aboard the space station, when we broke bread in fellowship with explorers from countries we'd previously regarded as enemies. Once again the potential of space travel as a unifying force was made wonderfully clear.

In September I attended an annual awards dinner and auction presented by Shades of Blue. The Colorado-based nonprofit encourages students to pursue opportunities in aviation and aerospace, among other STEM careers. The organization was founded by Willie Daniels who worked his way up from being a flight attendant to being a United Airlines pilot who flies Boeing's flagship 777. I was pleased to receive the Ed Dwight Jr. Education Award, named after the pioneering black astronaut. After leaving NASA, Dwight eventually returned to school and earned an MFA degree. He went on to become an eminently successful sculptor whose commissioned works include several acclaimed public installations around the country. In combining a life of science and art, he epitomizes the merits of STEAM, and I was proud to accept the award that bore his name.

The following month, I traveled to a continent I had only seen from space: Africa. I remember looking down on the continent, feeling awestruck by the light from some of the massive thunderstorms and the eerie darkness that covered remote areas like the Cape of Good Hope. On the ground in Cape Town, South Africa, I joined my colleagues participating in the International Space Education Board delegation to South Africa. Established in 2005, the board has a twofold mission: to promote STEM literacy and its importance to the space programs and the overall development of a well-educated and well-rounded workforce. Toward those ends, we met with the fledgling South African National Space Agency to bring them on board as a member. Our team also took books into the townships and built Lego-modeled solar cars that kids could take home and show their parents. The cars used the power of the sun to energize their small motors. Mabel Matthews and Carol Galica of NASA's Office of Educational Programs were instrumental in making this happen. I was amazed at the work and impact my team could make doing international diplomacy through our most precious resource, our children.

Our group had the opportunity to visit the Oprah Winfrey Leadership Academy for Girls in Meyerton Gauteng. I will never forget when a young student from Johannesburg asked me how she could work for NASA. I responded by saying, "Did you know there is an African Space Agency? You should look into how to be part of that, bring it to prominence, and be the first government astronaut to fly from Africa."

In truth, Mark Shuttleworth became the first South African to fly in space. However, he was a civilian who paid $30 million to fly as a tourist astronaut. It's the same whether you are a tourist astronaut or a highly trained astronaut. We both want to explore space and make life better for our civilization.

I also had the chance to visit the Cape of Good Hope within the Cape Point Nature Preserve, the continent's most southwestern point. I met Margie Kumalo at the board meeting, and she was kind enough to show me some of the sights. We rented a car and drove to the cape. From space, it looked like an eerie, dark void. Up close, it was a grass- and shrub-covered stretch of rocky terrain where the Atlantic and Indian Oceans meet. It's not every day that you can stroll in a park and see baboons and penguins.

The food was different, too. In Capetown, I dined on local delicacies, like gazelle, kudu, and other wild game. Near the cape, we stopped at shanties along the road to sample fresh crab, cod, and rock lobsters. Biltong, a beef jerky, and dried fruit are also popular and plentiful. At the Cape of Good Hope, I bumped into some German flight controllers who thanked me for providing them with their job security by safely installing the Columbus research laboratory during the *Atlantis* flight. It really is a small world after all.

A few weeks later, I joined the pop star will.i.am at Cape

Canaveral to watch the Curiosity rover take off for Mars. The $2.5 billion vehicle was about to undertake a two-year mission to search for signs that the Red Planet may have been habitable to microbial life. Will.i.am's fascination with space is well known, and Charlie Bolden, NASA's head honcho, had wanted to find a way for the agency to collaborate with him. Charlie put me in touch with will.i.am and we talked by telephone over the summer. We started brainstorming with a team including my friend Lars Perkins and we came up with a program called Stimulating Youth in STEM (SYSTEM).

Prior to the launch of the Curiosity space rover, we did interviews together with TV stations and networks throughout North America. I told a CNN anchor, "Maybe there's going to be a kid watching Curiosity traveling on the surface of Mars and they may think—well, how do I become a scientist or engineer? How do I become a musician? Because music and math use both the right and left side of the brain. So, if you know music, you know math, and let them see that—there's science and engineering in the arts too." At the launch, will.i.am and I wore T-shirts promoting the new venture. Afterward, he performed a concert for NASA employees that had us dancing into the wee hours. Will.i.am was so inspired by the experience that he went home and composed a song called "Reach for the Stars," an orchestral piece with a children's choir. When Curiosity landed on Mars, the Jet Propulsion Laboratory flight controllers sent a command to the rover on Mars to upload the tune. Once done, will.i.am's song found a new home on the Red Planet.

At home in DC, it sometimes seemed as if I had barely settled in when it was time to head out the door again. I was a fan of the open road and the friendly skies, so that was fine.

Still, I also enjoyed quiet evenings around the house, walking the dogs after work, making and eating delicacies that I treasured, such as split pea soup and salmon grilled with rosemary. I didn't even have to leave town for one of the year's final events. The U.S. Agency for International Development (USAID) had sponsored its "All Children Reading: A Grand Challenge for Development." Alex Trebek, the host of *Jeopardy!*, emceed the event. It took place in the Ronald Reagan Building, a few blocks from the White House. About eight hundred people were on hand.

According to the program's premise, 171 million people could be lifted out of poverty if all students in low-income countries left school with basic reading skills—equivalent to a 12 percent cut in global poverty. To improve student reading outcomes and access to future economic opportunities, USAID and its partners announced a $20 million effort to promote literacy among children in the primary grades around the world. The campaign dovetailed nicely with our own efforts at NASA Education and I was happy to attend and lend my support, along with Secretary of Education Arne Duncan and other dignitaries engaged in the battle to improve learning skills. U.S. Representative Nita Lowey was also in attendance. "Literacy is one basic skill that opens door after door," she told the audience. "Reading is the pathway to education, and education is a cornerstone of free and stable societies." Remembering my childhood bedtimes when my mother led me through the pages of *The Little Engine That Could* and *Curious George*, I knew that reading could also be the pathway to the stars.

"All Children Reading" wrapped up a season of transition, during which I traded the blue suit of the astronaut program for the suit-and-tie uniform of an agency leader. The change, while comprising the challenges and surprising twists that any transformation involves, had still taken place fairly rapidly. I had been back from space for a year.

• • •

I began 2012 with more activities designed to introduce young learners to the kind of academic disciplines that could lead to careers in space exploration. Among the most rewarding of them was the Symposium on Supporting Underrepresented Minority Males in Science, Technology, Engineering, and Mathematics, which we sponsored in collaboration with my friend Dr. Lorenzo Esters from the Association of Public Land-grant Universities (APLU) and the American Association for the Advancement of Science (AAS). Our day-long conference coincided with the publication of an APLU report, based on a recent survey of minority students, higher education faculty, and administrators, that drew a sharp set of conclusions: To succeed in science-related studies and professions, motivated men from underrepresented minority groups need active engagement and mentoring by college faculty, personal involvement in undergraduate research, and adequate financial support. I had benefitted from the kind of engagement described in the report, and I pointed out as much in my opening remarks at NASA HQ in Washington. I credited my success to the village that nurtured me, including Woodrow Whitlow, my longtime colleague at NASA, and Charlie Bolden. They were two minority male mentors who impacted me at an early age and helped me get to where I needed to go.

"Our vision is to advance high-quality science, technology, engineering, and mathematics using our unique capabilities at NASA," I continued. "We recognize there's a gap that must be bridged as it relates to minority males in STEM. We identified the challenges and best practices. Now it's time to go the next step and really have a call to action to increase minority participation in STEM fields."

I conveyed a similar message while speaking to educators at the National Science Teachers Association's 60th National Conference on Science Education in late March. I was on a panel with Anousheh Ansari, the first Iranian woman to fly in space as a tourist. She paid $30 million to live on the ISS for a week. Anousheh shared her experience of immigrating to America, obtaining student loans, and working hard to get through college while learning English in the process. She started companies in the tech sector and was able to bankroll her ride to the cosmos. Like Venus Williams's inspiring story, Anousheh's tale was yet another demonstration of the value of grit and dogged perseverance. I collected such stories as I traveled and worked with successful people from around the world, intent on using their examples to motivate the next generation.

I recalled Anousheh and all the other exemplary people I met through my involvement in the space program later that summer, when my education team and I walked from our offices on E Street SW to the reflecting pond in front of the U.S. Capitol building, our eyes on the skies. The Shuttle Carrier Aircraft (SCA) did a fly-by before heading to the Udvar-Hazy Center in Virginia, where space shuttle *Discovery* would be displayed for all eternity. I watched history being made while snapping the shutter furiously on my camera. As *Discovery* rode piggyback on the SCA, I was able to capture an image of the two vehicles perfectly aligning with the spire of the

Capitol, where John F. Kennedy had petitioned Congress for resources to fund the space program more than a half century before. We later had a handover service at the museum, with more than sixty astronauts and thousands of space enthusiasts bearing witness. Charlie Bolden, who had flown two of his four missions on *Discovery*, signed over our bird to the Smithsonian and told them to take good care of her. I remember taking pictures in front of *Discovery* with many of my classmates, including Paolo Nespoli, Christopher Ferguson ("Fergy"), and Dex.

The Summer of Innovation was going full steam by July, and so was I. Traveling around the country to promote STEM, I constantly encountered teachers and students who were excited about NASA, and they had questions about life in space. One popular question is, "What does it feel like when you first get into space?" That's easy to answer. Once you're free of Earth's gravity, a lot of interesting things happen. You first notice the small items that are unattached to anything in the shuttle start to float. A pen you may have dropped or dust particles—all slowly floating upward although you're still strapped into your seat. The fun really begins once you're free of your seat. Once unstrapped, your body floats, too. Astronauts orbiting the Earth are subject to microacceleration. Think of it this way: Once you are free of gravity, if you bump into anything inside the shuttle, you'll simply bounce, much like a rubber ball thrown against a wall.

There's another condition astronauts experience even before they can unstrap themselves—the sensation of tumbling. It occurs once the main engines shut off. Imagine if you were driving and suddenly hit your brakes. You'd lurch forward.

It's a similar but a more intense reaction in space that starts right after the main engines stop. The rate of acceleration changes, but since there's no gravity, your inner ear and brain react differently. You feel like you're tumbling, still moving forward in a state of motion. All I can say is thank goodness astronauts are trained to function in gravity-free conditions.

Every now and then I would get some questions that might appear odd or sound downright strange. I always try to keep an open mind about it because very few people on the planet have ever traveled in space. A student at an international school in Washington, DC, once asked me if I had ever encountered any "green aliens with pink lipstick." She was about six years old and eagerly awaited my response amid all the laughter and snickering. "I haven't seen any aliens yet," I said. "But that doesn't mean they're not out there. That's why we continue to explore." One question I will never forget came from a teacher who pulled me aside after a presentation at the Kennedy Space Center. She seemed nervous and admitted to being a little embarrassed to ask. But she pressed on: "Is sex possible in space, and can a man get 'hard' up there?"

I paused, searching for the right words and an honest answer.

"Blood flows on the ground," I said. "Blood flows in space."

One additional benefit of my work was the chance to spend time with my friends from the Corps. I connected with Bobby Satcher in DC, where he was preparing to leave NASA and return to practicing orthopedic surgery. We reflected on the good ol' days in space and shared our concerns about providing opportunities in STEM for underserved populations. I also got to see my old friend Charlie Camarda in New York.

We were both on hand when the *Enterprise*, NASA's very first space shuttle, joined the Intrepid Sea, Air & Space Museum's permanent collection.

In August, I rejoined will.i.am at the Jet Propulsion Laboratory in Pasadena, when "Reach for the Stars" was beamed via radio signal about 150 million miles back to Earth from Curiosity's landing spot on Mars's Gale Crater. Students from the East Los Angeles neighborhood where will.i.am grew up were on hand at the propulsion lab to witness the historic event. He hoped that enabling the students to see that epic marriage of science and music would inspire them to take the song's message seriously. The point is to remind people that anything is possible, he said, "If you discipline yourself and dedicate yourself and stand for something. We don't have to just end up in the hood. But it's a hard thing. The hardest thing is discipline."

I echoed his remarks. "Never give up," I told them. "People told me that I couldn't be an astronaut. Whatever you want to be, whatever you dream, you can do, if you put your mind to it."

That fall, the man who truly embodied space exploration passed away. Neil Armstrong, who took that first giant leap for mankind on the moon in 1969, died from complications following cardiovascular procedures. Armstrong exemplified a humble, workmanlike attitude. He was kind and shy, very modest and understated in his bearing. He was also a legend, a larger-than-life figure, but I had the impression that he didn't want to be perceived that way. Instead, he considered himself to be a practical, functioning engineer who focused on his job. Despite his modesty, Armstrong had become the face of NASA in the early days of lunar expeditions. His public memorial service was held at the National Cathedral in Washington, and I was proud to be on hand with most of the Astronaut Corps who knew and worked with Armstrong, men like John Glenn, Gene Cernan, Buzz Aldrin, and Michael Collins.

"Fate looked down kindly on us when she chose Neil to be the first to venture to another world and to have the opportunity to look back from space at the beauty of our own," Cernan said during his eulogy. "No one, but no one, no one would have accepted the responsibility of his remarkable accomplishment with more dignity and more grace than Neil Armstrong. He embodied all that is good and all that is great about America."

On a personal level, no one embodied the best of our country as well as my father. Every lesson he taught me in some way pointed me toward my possibilities. When I visited my parents in Lynchburg that October, I could see that Dad was slowing down considerably.

The stroke and ministroke that he had suffered before my last shuttle mission were starting to present some persistent medical issues. Looking at him, I recalled the times when he would snatch me up into the air with one arm and sling me into the water while camping in Cape May, New Jersey, in the summers. This once-powerful athlete, musician, and educator was starting to slump over in his chair after a meal or lose his train of thought as a mild form of dementia started to take hold. Cathy and I were starting to think about the challenges of caring for our aging parents. We had moved them from the Cape Cod on Hilltop Drive because the multiple flights of stairs had become difficult to navigate. Their new home, a single-story condo, was much easier for them to move around in. Like so many other Americans of our generation, Cathy and I were experiencing a role reversal in which we had become responsible for our parents.

Still wrestling with those changes at home, I headed off to Naples, Italy, for my second International Space Education Board meeting at the International Astronautical Congress. Once again we worked with students and educators to bring space down to everyone. Space agencies representing Japan, South Korea, Europe, Australia, South Africa, and Canada were all working together for the sake of education. I again saw the flight controllers from Germany that I had seen in South Africa the year before. Jovial and familiar with each other, we posed for another picture commemorating the installation of Columbus. For our cultural excursion we took the students to Pompeii and looked at the fossilized devastation resulting from a mountain blowing its top. Seeing the historic aftermath of these interactions between nature and humans made me think of the really big explosions from our solar system forming and giving birth to stars, cosmic collisions spawning galaxies. Articles of faith and scientific concepts swirled in my mind like electrons circling a nucleus. Big bang. Creation. Fission and fusion. Intelligent design. I meditated on the complex relationship between the macro and the micro, the various ways in which one form of something can lead to the other. Small building blocks creating bigger blocks. Atoms blooming into atmosphere. Carbon, hydrogen, and oxygen combining to form life. How each of us is so wonderfully made.

Our faith contends that life continues in the hereafter after we have breathed our last breath. Similarly, science observes that energy is neither created nor destroyed but simply takes another form. Both notions provided comfort in November, when we honored the memory of my Penguin classmate Alan "Dex" Poindexter. An accident while on vacation had taken him from us, and we laid him to rest at Arlington National Cemetery on a crisp, sunny day. In addition to joining the Corps together in 1998, Dex and I had flown together on STS-122. His joyful spirit and selfless compassion had helped to make that mission a success, and all those who knew and loved him would sorely miss him.

After the ceremony, I had received a call from my wonderful neighbor Melissa that my dog Scout was not doing well. He had been diagnosed with lymphoma and had been treated with steroids. When I arrived home that night, I had to rush him to the hospital because he was not responsive. That same night I put my boy down, bringing our wanderings as a trio to a bittersweet end. Scout had been a kind, loving dog that became instant friends with anyone he met, and his death punctuated a season of loss. With Jake at my side, I returned to Lynchburg for Christmas and bolstered my spirits through bonding with friends and family.

• • •

Rested and renewed, I started off the new year with a speech at a Martin Luther King Jr. Day observance cosponsored by the Roanoke Chapter of the Southern Christian Leadership Conference (SCLC) and the local branch of the NAACP. My talk enabled me to reconnect with my Virginia roots while also helping black Virginians visualize the possibilities for our children. I remember speaking in front of a banner exhorting viewers to "Move forward in difficult times" and reminding them of the Reverend King's famous maxim: "Injustice anywhere is a threat to justice everywhere." To illustrate my point about equal opportunity, I showed a slide of three boys of varying heights trying to watch a baseball game from behind a fence. The first panel, above a caption saying, "This is Equality," shows the boys standing on boxes of the same height. The two taller boys can see over the fence while the third boy is forced to peer between slats in the fence. The second panel is above a caption that says, "This is Justice." The illustration shows the two shorter boys comfortably viewing the game from boxes custom-designed for their individual statures, while the tallest boy is able to watch without the benefit of any box at all. The caption above both panels proclaims, "Equality doesn't mean Justice." I stressed that our children needed more than just access to the benefits that STEAM education offered; they also needed the necessary support systems to ensure that access leads to success.

Later that month, I got to witness the epitome of black success in America at President Barack Obama's second inauguration. I was part of a group of NASA engineers and astronauts accompanying exhibits of the Curiosity rover and a float of the Tuskegee Airmen. When we got close to the presidential viewing area, I saluted the president. He looked my way and saluted me back. Once again, seeing President Obama take the oath of office filled me with immeasurable pride.

As part of the inaugural festivities, I spoke to kids at an event sponsored by Jack and Jill, the African American family organization. After that I went to a party with Pharrell and will.i.am to look at using gaming for education, part of NASA and White House efforts to engage nontraditional partners to help advance the STEAM cause.

The following month will.i.am invited me to attend his TRANS4M 2013 at the California Science Center. The focus of the event centered on wellness and STEAM. *Los Angeles* magazine described the conference as a place "where geek meets cool." In addition to presentations from me and science luminaries such as inventor Dean Kamen, the creator of the Segway, the conference showcased youngsters from Los Angeles' Boyle Heights area, will.i.am's old neighborhood. Those kids are doing amazing things with science. Many of them were students from Roosevelt High School, depicted as a "Drop Out Factory" in the award-winning documentary *Waiting for Superman*. Will.i.am said Superman was never going to come to his neighborhood and therefore he needed to put on the cape himself, with help from corporate and local support. I began the day's proceedings while standing under the space shuttle *Endeavour*, and former president Bill Clinton delivered the much-anticipated keynote. While stressing the importance of the arts and sciences, he suggested that youngsters growing up in tough neighborhoods could also be especially equipped to succeed.

"A lot of them are unusually alert to their surroundings at a very young age," Clinton said. "The observational powers they need to survive on the streets are the same they need to apply in class." Will.i.am closed the conference by reiterating the goals for TRANS4M: "Kids are the architects of the future. We need to give them the tools to design their tomorrow to be different than their predetermined future." Afterward, Alicia Keys and the Black Eyed Peas jammed at the Hollywood Hotel late into the night.

The rare opportunity to party with the stars and other

luminaries brought levity to the workday grind of managing budgets and navigating office politics. Even all that pales in comparison to the byzantine workings of Capitol Hill. In June, I had to testify before the House Committee on Science, Space, and Technology. I had a dual role, speaking as both a senior NASA official and a cochair of a White House committee called the Federal Coordination in STEM Education Task Force. Budget hearings typically pit different government agencies and special interests against each other, jockeying to control programs and spend taxpayers' money. Sitting next to my cochair Joan Ferreni Mundi, who was also the education director at the National Science Foundation, I had my prepared remarks describing our plan to best utilize the $2 billion in the federal budget earmarked for STEM education programs.

I told the committee, "For the United States to maintain its preeminent position in the world it will be essential that the nation continues to lead in STEM, but evidence indicates that current educational pathways are not leading to a sufficiently large and well-trained STEM workforce to achieve this goal. Nor is the U.S. education system cultivating a culture of STEM necessary for a STEM-literate public. Thus it is essential that the United States enhance U.S. students' engagement in STEM disciplines and inspire and equip many more students to excel in STEM." I went on to outline our framework for increased collaboration among agencies and how to strengthen partnerships with other agencies to provide a coherent and cohesive network of STEM education efforts at the federal, state, and local levels.

After my testimony, the committee began questioning me about my involvement in NASA budget cuts to seventy-seven of its educational programs. Donna Edwards, a Democratic congresswoman from Maryland, asked me point-blank if I had cut my programs. I hesitated before finally responding no, which prompted the congresswoman to ask the question again. There was a reason for the pause. Before the hearings, my boss and NASA's deputy administrator had urged me to tell the committee that I had cut the programs. I told her I couldn't do that because it wasn't true. She chewed me out after the hearing. Fortunately, the president's science adviser sent me a letter of appreciation for doing a great job on the Hill. I felt vindicated.

Two weeks later I attended an astronaut reunion in Houston that enabled me to reflect fully on the rich history of black members of the Corps and my place as a grateful beneficiary of their legacy. I met with Cheryl McNair, widow of Ron McNair, who had died on the shuttle *Challenger*. Legend has it that when he was finishing his PhD in laser physics at MIT, someone stole the briefcase that contained his almost-completed dissertation. Back then it was all done on typewriter without the backup files we rely on today, so he had to reconstitute everything from memory to graduate. A brilliant black thinker, he grew up in South Carolina, where someone called the police when young Ron tried to check out a science book from the segregated library in his town. Once the police officer arrived, he saw Ron's determination to learn and was so moved that he checked out the book for Ron.

A role model to me even though I never met him, Ron had convinced Charlie Bolden to apply to the Corps, who in turn told me that you can't get in unless you apply. Theirs was a legacy of encouragement passed down to each successive generation of black astronauts, a tradition that I worked to extend by helping a new generation of black students pursue STEAM educations.

Ron had been the second black American citizen to fly in space. The first, Guion Bluford, was also at the reunion. Understated and approachable despite his formidable role in history, he was always quick-witted and sharp as a tack. In addition to spending time with pioneers like Guion, I enjoyed the added bonus of catching up with my Penguin classmates.

My fellow astronauts formed a valuable kind of family, strengthened by relationships built over time and sustained by shared experience. That summer I attended another reunion, a gathering of the family that I was so blessed to be born into. My mother's side of the family, the Colemans, gathered in Halifax, Virginia, in July. My grandparents had a tobacco farm and my mom had always told me that the work was hard. They had cows, pigs, chickens, and the usual farm animals that made life manageable without much outside intervention. Uncle Chandler still runs the farm today primarily by giving horse rides, and makes his living as an auctioneer. I had never heard someone speak as fast as he did when soliciting bids on items for sale.

Listening to my aunts and cousins recalling great stories from days gone by, I traveled in my mind back to my own childhood experiences at our family farm in Halifax, Virginia. One summer I decided to venture through the tall grass in my plaid shorts and Buster Brown shoes. Earlier in childhood I once wore braces because my legs were bowed—a trait I inherited from my dad. However, I had outgrown the need for metal support and was just happy to walk through the fields on my own. I was unaware though of what was in the tall grass as I returned home covered from ankle to knee in chiggers. The little mites had burrowed under my skin, and I soon started scratching. I couldn't stop. My grandmother coated my legs with calamine lotion, but that only gave me temporary relief. She then proceeded to paint my legs with white shoe polish. She said the polish would harden and suffocate the bugs. I was a sight to be seen, walking around the yard with little white legs sharply contrasting with my brown skin. My sister, Cathy, and my cousin Karen could not stop laughing and picking at me, but I didn't care because Grandma's home remedy was indeed easing my anguish.

At the reunion, we had our customary talent show with the kids, along with lots of eating and dancing, all of us clad in white linen, clapping and whirling in rhythm. My dad even got on the dance floor, leaning on his walker and grooving to the beat of Michael Jackson's "Rock with You." As I look back on it now, a Luther Vandross song comes to mind. In "Dance with my Father" he says he'd play a song that never ends if "he could get another chance, another walk, another dance" with the man who guided him to maturity.

That song easily conjures up all the charm and comfort of home but returning to one's roots isn't just about nostalgia. It's also about responsibility, the desire to do something meaningful for those who have given you so much. Those impulses were among the many thoughts consuming me when I returned to Virginia for Thanksgiving. In between trips, I had been to San Francisco, China, the University of Maryland, and elsewhere, promoting science and the value of scientific discovery. In Lynchburg I saw an opportunity to reflect on the value of family ties. I ran into my old friend, Ralph "Chopper" Wilson, who was a barber and now real estate developer building luxury condos in downtown Lynchburg. He's a kid from the hood done good. His grit and determination had paid off, big. I ended up negotiating with him and renting a new condo to spend more time at home.

Back in DC, Jake, my running partner of fifteen years, was in rapid decline, his mind and body worn out. Before Jake died, we spent the day at the Arboretum, listening to the sounds of the city. The next morning he was gone. Soon after, I began to pack my things. Although I hadn't worked out the details, Jake's passing had confirmed an impulse that had been growing in my consciousness. It was time to go home.

• • •

"I am sorry to inform the NASA family that my good friend and our associate administrator for education, Leland Melvin, has decided to retire next month after more than twenty-four years of NASA service," Charlie Bolden wrote in a January memo to my colleagues. Twenty-four years working with a single organization is almost unheard of in this day and time. There were so many jobs I had done within NASA. I had worked as a research scientist at Langley, writing papers and presenting my fiber optics and optical NDE (Nondestructive Evaluation) work at conferences around the country. I had been a student when I went back to school to work on a PhD at University of Maryland in mechanical engineering. I had been a project manager that helped lead a team to create the first ever in-line optical fiber sensor manufacturing system. I had been an astronaut; I had been an educator; I had been a senior executive service member working for Administrator

Charlie Bolden. NASA provided an incredible journey that allowed me to learn so many different things, meet people from all over the world, and leave the planet twice. As I prepared to leave, I remembered Rosa Webster, who saw something special in me and recruited me to join the agency. I thought of Katherine Johnson, Woodrow Whitlow, Bob Lee, Bill Prosser, Robert Rogowski, Tom Kashangaki, and all the other people who made sure I understood the possibilities and never talked about limitations. I left with the same impressions I had when I began: NASA is a combination of many things but most of all it's a family.

I returned home on February 16 and had a wonderful conversation with my dad. As we talked about his basic needs, I understood that our roles had all but completely reversed. He also talked about his beautiful wife taking care of him and how he felt bad about being unable to take care of her. After a fruitful discussion, I helped my parents to bed and headed downtown. The next day, my sister, Cathy, drove him to rehabilitation therapy and I joined them there. After the session, I was following them in my car when I noticed that Cathy had pulled over. My dad was slumped down in the passenger seat, unmoving. We rushed him to the emergency room, which was not far away. The doctors tried to revive him to no avail. As our friends started to hear the news of my father's passing, the hospital room started to fill with familiar faces. It quickly turned from being a place of sadness to a place of gratitude for the impact that he had made to society. People started telling stories about him and we prayed and sang in his name. It became a very spirit-filled place, full of joy. I had very mixed emotions because I had moved home to be there for him. I had hoped to somehow bring things back to the way they were, with me being his son and he being my guide, my mentor, my dad.

• • •

It was once hard for me to wrap my head around intercontinental travel because as a kid I had never flown in an airplane. All of our family travel was either in an RV or by Greyhound bus, and sometimes the train. But to think about getting to the places that I had only looked at in my encyclopedias and maps was something I could hardly fathom. I slowly developed an understanding of the larger world through my father's patient tutelage. He had introduced me to the principles of engineering when we converted a bread truck to a camper, little knowing that our labors would someday lead to my work in perfecting optical sensors for space vehicles. Just as the view from space illuminates connections on Earth that are seldom seen at ground level, my perspective as an adult helped me see the links between my father's teaching and NASA's mission. The NASA Space Act of 1958 required the agency to provide "the widest practicable and appropriate dissemination of information concerning its activities and the results thereof." To me this meant education and inspiration, in the same way that my father had educated and inspired me. At NASA we worked to let kids, parents, and teachers know what we were doing and why. We attempted to show how everything was connected as one huge planetary ecosystem, solar system ecosystem, and universal ecosystem. Charlie Bolden's memo about my retirement had also said of my work: "Using NASA's unique missions, programs, and other agency assets, he has helped cultivate the next generation of explorers—one that is truly inclusive and properly reflects the diverse makeup and talent of this nation's youth and our agency's future." I knew that wherever my own future led me, my efforts would similarly involve spreading the word about possibilities and associations, lighting the path for journeys humanity has yet to take.

11

The Next Mission

T he ache for home lives in all of us," Maya Angelou once wrote, "the safe place where we can go as we are and not be questioned." Back in Lynchburg, I reconnected with places and people whose presence fulfilled that ache in me. Being home gave me a feeling of acceptance that other places can seldom equal.

I reconnected with Ernest "Fufu" Penn, one of my old friends and charter member of the Big Blue Crew. Our dads had coached our baseball team when we were in middle school. One year, when we were both selected to compete in the All-Star game, we all drove to the championship in Winchester, Virginia, in my father's bread-truck camper. Now an ordained minister at Mount Sinai Baptist Church, Fufu came by my parents' home after Dad's funeral. He wanted to share with me something he had been praying about. Fufu told me that I thought I had come home for my father, but I had really returned to release him by being available to help take care of my mother. My dad, knowing that his queen was going to be taken care of, did not have to worry about leaving this planet for a more peaceful place, heaven. I needed to hear Fufu's compassionate words. They really helped me through the grieving process.

I was also able to get together with Stan Hull, another old friend who had played a pivotal role in my life. Now an assistant principal at Brookville Elementary School, he had been the quarterback who threw me the winning touchdown pass. I visited him at his school and met some of the science teachers there. I told them if it hadn't been for Stan, I wouldn't have been there to talk with them. Although I was joking, there was some truth to that statement. His perfectly thrown pass into the end zone had unleashed an incredible series of events that placed me on my current path.

In March that path took an unanticipated turn when Creative Artists Agency (CAA) asked if I would be interested in joining their roster of public speakers. I had met two CAA agents, Darnell Strom and Michelle Kydd Lee, about six months before at a STEAM program at the CAA headquarters in Los Angeles. At the time, I had been working with Connie Yowell, the director of education at the MacArthur Foundation, traveling with her team around the country to promote opportunities for kids to learn outside of school. We both believed it extremely important to go where kids are and use their experiences to create STEAM teaching moments. During our tag-team appearances, I told my story and Connie shared the gospel of connected learning. After I retired from NASA,

CAA called me and soon I found myself among the ranks of such luminaries as Will Smith, Bishop T. D. Jakes, and Mike Ditka, to name a few. Imagine the most dominating talent agency selecting me to do pretty much what I enjoyed doing at NASA?

I was now a motivational speaker and really did not know what that would mean or how successful I would be. I had my NASA pension to support me if things went sideways, and the cost of living in Lynchburg had become manageable. I had given hundreds of talks for NASA but now I was pitching my own story without the NASA marketing machine and one of the most trusted and well-known brands, the NASA logo. Seventeen years after adopting a minimalist design that spelled out the agency's name in serpentine letters (earning the nickname "the worm"), NASA reinstated a logo known as "the meatball." The latter features a blue planet speckled with stars, adorned with a red chevron (representing wings), and the circular flourish of an orbiting spacecraft. Even in the midst of shifting insignias, NASA's brand continued to suggest innovation, vision, and exploration. Could I transform myself with similar success? Only time would tell.

May was very busy.

Lynchburg is roughly 180 miles from Washington, DC, but in some respects it could be on a different planet, especially when it comes to the corridors of power. I got to see the differences firsthand when I visited the White House. Just six years before, I had watched proudly as Barack Obama took the oath of office. Now I was in the Obamas' house, at their invitation. As a guest participant at the annual Easter Egg Roll, I read *The Little Engine That Could* to a diverse group of children on the White House lawn. Afterward I took a photo with the

First Couple, who couldn't have been more gracious. Almost immediately afterward, I received a kind letter from Michelle thanking me for my participation. It wasn't the first time I got a letter from the Obama White House. In April the president had written to congratulate me on my twenty-four years of federal service. "Public service is an honorable calling," he wrote, "and it is my privilege to join in celebrating your career."

It was also my honor to receive recognition from Congress. Virginia's Senator Tim Kaine, best known as Hillary Clinton's running mate in the 2016 presidential election, read a proclamation on the Senate floor. "I commend Leland for his commitment to science and education, as well as public service," he read. "At a time when STEM education is becoming a priority for the United States, Leland's work has been beneficial to developing the skilled workforce necessary to drive our nation's world-class innovations."

In May 2010, Bobby Satcher and I had been awarded honorary doctorates at St. Paul's College, a historically black college in Lawrenceville, Virginia. That moment came to mind as I thought about my future work. The school had special significance for both of us. My father had graduated from St. Paul's, and Bobby's father, Dr. Robert Satcher Sr., was president of the college. Bobby's uncle David was another notable Satcher. He served as U.S. surgeon general under President Bill Clinton. The college also honored Tom Joyner, the radio host with whom I had last talked while circling the planet. Tom had made his reputation as "The Fly Jock" in the late 1980s, when he hosted daily shows in Chicago and Dallas. He broadcast in Dallas from 5:30 a.m. to 9:00 a.m. Once finished, he hopped on a plane to Chicago where he hosted another radio show from 2:00 p.m. to 6:00 p.m. Both broadcasts were the top-rated radio shows in their respective time slots. His tireless approach and determined work ethic provided a valuable example to young people everywhere. Over time, he became as recognized for his philanthropy as his skills in the radio booth. At St. Paul, he gave five dollars to every graduate and we took pictures with each of them. It was a special moment for us all, and as we watched the graduates stroll off into the future, we reveled in the certainty that some among them would save lives, develop cures, change the world. Now I was retiring from NASA but not retiring from my work.

My friends and colleagues hosted my official NASA going-away party on May 17. At the Doubletree Hilton in Crystal City, Virginia, about two hundred people gathered in a spacious ballroom. Elaborate assemblies of plates and flatware adorned each table. The centerpieces, clear vases about three feet tall, held glorious sprays of light blue, dark blue, and yellow flowers. A small candle flickered at the base of each vase. A large sheet cake featured my NASA photo with Jake and Scout. Another cake featured a portrait of me in suit and tie.

Alan Ladwig, the former deputy associate administrator for public outreach at NASA, served as the master of ceremonies. Quick-witted and comfortable with a microphone, he gave the party a festive, irreverent air, half tribute, half roast. Before his retirement, he had been instrumental in the space program in a number of ways. He had started the Teacher-in-Space program and had been on the selection committee that chose me to be associate administrator for education. At the party, he presided over a succession of featured speakers who had known me at various stages in my life. People like my high school buddy "Silky Blue" and my college coach Morgan

Hout. Along with Allyn Griffin, who had played with me on the Detroit Lions, they covered the early, pre-NASA years. Dr. Myers, who had declined the honor council's request and helped me back on track at Richmond, recalled my time on campus. Woodrow Whitlow, who had been a colleague for so long, addressed my early career at NASA. Listening to him, I fondly recalled the time he threw a party for me when I got accepted into the Astronaut Corps.

Charlie Bolden couldn't attend but his family came in his stead. His granddaughters bestowed me with commemorative NASA pins after he delivered a special message via video, saluting my "incredible career" and calling me the most unique individual he'd ever had a chance to meet. "Thank you from the bottom of my heart," he concluded, "for all that you have done not just for NASA but for the nation and for young people with whom you've come in contact and whose lives you've touched in your own very special way."

We also heard video testimonies from people I'd known in recent years, entertainers and artists like Arsenio Hall and Quincy Jones. "I will never forget the time we spent together with Pharrell that afternoon at the space museum in Washington," Quincy recalled. "I wish you everything that you wish for yourself." Pharrell, whose parents were on hand, was unable to join us. He also sent videotaped greetings from out of town.

Of all these wonderful testimonies, two of the most unique came from Christian McBride and Laura Rochon. A Grammy-winning virtuoso, Christian played "Fly Me to the Moon" on the acoustic bass in his inimitable style, taking me totally by surprise. Laura, a public affairs specialist at NASA, also does an excellent Sarah Palin imitation. For her video, she dressed up as the former vice presidential candidate and offered a farewell in the form of a campaign commercial. While pretending to be unable to tell the difference between stem-cell research and STEM education, she shared details from our friendship and my career. "I think I can speak for many here tonight by saying we celebrate not just Leland's achievements but his character," she declared, sounding remarkably like Palin. At the end, she unveiled an election placard revealing the "real" reason for my retirement: a Palin/Melvin presidential ticket.

The most poignant moment belonged to Ed Dwight, an Air Force employee in the early days of the space program. Ed wanted to become an astronaut but faced hostilities and hurdles because of his race. He didn't get that far, thanks to a brutal system of racial discrimination. People in the audience gasped audibly as he described the barriers. If circumstances had been different—indeed, if the country had been ready for him—Ed most certainly would have become the first black astronaut in space.

My friend Simon Sinek, the inspirational speaker and NASA consultant, offered a champagne toast. His words described the agency's mission from the days of Ed Dwight to the recent shuttle voyages to the International Space Station: "Reach for new heights and reveal the unknown so that what we do and learn will benefit all humankind." They were my aspirations as well.

The party was an incredible way to close out my career with friends and my NASA family. That so many people—from my childhood throughout my entire career—thought enough to come and honor me in such a beautiful way really touched my heart.

I tried to capture all the emotions I was feeling. When I finally did speak, it was one of those occasions where language falls short despite our best efforts. "I am so honored to have been able to work with so many incredible people throughout my journey," I said. "Unfortunately my father couldn't be here to see the culmination of what he and my mom produced on Pierce Street and later on Hilltop Drive. I know he's up there looking down and smiling. It's all kind of full circle. Through all of your efforts, having my back and believing in me, he's proud."

• • •

By that point in time I had spoken to audiences all over the globe, testified before Congressional committees, conferred with world leaders, and answered interviewers' questions while working in space. But I had never quizzed brilliant youngsters competing for a lucrative prize. As it happens, that's what I wound up doing next.

Darnell called me from CAA and said a producer, Laurie Girion of Shed Media, wanted to talk to me about being the host of a television show called *Child Genius*. It had originated in the United Kingdom and was being adapted for a U.S. audience. Laurie and I Skyped that afternoon and I shared my story with her. She seemed impressed. We Skyped again the next day so they could record the session for the head of the Lifetime Movie Network to approve. The interview went well and a few months later I was in LA at the Skirball Cultural Center meeting incredibly bright eight- to twelve-year-olds. The kids were funny and precocious. They were also serious because they all had a desire to win the trophy and the prize, $100,000 in college scholarships.

Lifetime's promotional description of the show outlined the challenges the contestants would have to surmount: "*Child Genius* centers on America's most extraordinary and gifted children and their families as they prepare for a national intelligence competition. In cooperation with American Mensa, the competition takes place over eight weeks and tests the nation's brightest young minds on their knowledge of Math, Spelling, Geography, Memory, the Human Body, U.S. Presidents, Vocabulary, Current Events, Zoology, Astronomy and Space, Inventions, Literature and the Arts, Earth Science, and Logic."

Over the course of the summer, we shot eight episodes. We would bring all the kids out to the Skirball Cultural Center in Los Angeles and film two episodes over four days. The production team created the set and brought everything to the cultural center. There was no rehearsal because it was a reality show and they wanted it as raw as possible, but I did study the questions and pronunciation of the extremely crazy words like "floccinaucinihilipilification," a multisyllabic word that means "the action or habit of viewing something as worthless." This was just one example of the vocabulary of these exceptionally talented kids.

I had to be very careful not to show favoritism. I also had to ensure that my response to questions was the same pace for all kids, to avoid protests from parents watching every little thing for signs of an unfair advantage.

Hosting the show introduced me to the terms "helicopter parents" and "tiger moms." Helicopter parents hover over their children, seemingly taking an excessive interest in every aspect of their daily experience. "Tiger mom" is a phrase popularized by Amy Chua in her book *Battle Hymn of the Tiger*

Mother. Chua, a lawyer and mother of two girls, has written, "I do believe that we in America can ask more of children than we typically do, and they will not only respond to the challenge, but thrive. I think we should assume strength in our children, not weakness." The parents I encountered during the taping of the show might not have described themselves as "tigers," but many clearly believed in demanding high achievement from their children. This was compelling to me because it contrasted with the methods of my own mom and dad, who never hovered. They just asked me if I had what I needed to get my work done.

Lifetime was aware of the drama lurking in those family relationships. One commercial proclaimed, "Behind every child genius there's a driven parent determined to maximize their child's potential." The producers kept me isolated from the kids and parents during the off-camera times so there would be no perceived inappropriate relationships. In addition, we had a game compliance person there to make sure nothing was fixed and there were no irregularities because there was $100,000 at stake.

I was nervous when I started because I had never done anything like that before, and after about three minutes on set a bead of sweat would pop on my head and the makeup person would have to come out and pat me down with more powder.

The Shed Media producer was in charge and had the last say on everything. She would sometimes come out and coach me on things they were looking for, or suggest that I say things with a certain inflection. We often did multiple takes of me asking questions like: "Take the number of ribs in the average human and multiply that figure by the two numbers representing Grover Cleveland's presidencies." Give yourself a pat on the back if you knew people typically have twelve ribs and Cleveland was the twenty-second and twenty-fourth president and came up with 6,336.

The extra takes enabled them to insert corrective material if I made a mistake. Of course, my goal was to do it in one take, but that didn't always happen. I'd flub some lines, mispronounce words, or botch a name, and we'd have to shoot the line over again. It was like training in the simulator. You just kept doing it until you got it right.

The studio audience was composed of parents, siblings, family members, and paid fill-ins. I learned to my astonishment that there is an entire industry of people getting paid to go from show to show and sit in the audience. The set had to be deconstructed every week because the location was only rented for the four days. I purchased my own suits but the producers had to approve my shirt and tie combinations before I wore them on camera. We also had to do regular texture tests to avoid the moiré effect in which the high-resolution cameras can make tight, overlapping patterns appear to vibrate.

The work from that summer aired in January 2015. We recorded a second season during the summer of 2015 to air January 2016. I had enjoyed a good measure of public recognition before *Child Genius*, but being on the show expanded people's awareness of me even more. Folks really loved the show because it was a positive program that celebrated brainpower instead of brawn or crazy behavior. In the end, however, it seems that more Americans were more interested in seeing Honey Boo Boos and dance moms parade across their screens, and we were canceled after two seasons. Still, I was grateful for the experience and the exposure. I could hardly believe that I had landed a show on national television so soon after I retired from NASA, something I hadn't expected at all. Once again my journey had confirmed the importance of having a core set of tools to do the job, even if you have no idea where you're going to end up and what tasks you'll be required to perform.

Meanwhile, my travels on the lecture circuit also reached liftoff. Over the course of 2014 I did six talks and a couple of cool digital campaigns for *The Verge, Flipboard,* and Avis. Each of those projects enabled me to use the Internet to introduce my story to a wider audience.

For *The Verge,* a net-based technology news and media network, I recorded a segment as part of its Chronicles series. It started with images from my youth, including childhood pictures and a photo of me at the high school homecoming game, chasing the touchdown pass, arms outstretched. In my narration I paid tribute to my parents' invaluable guidance. "I give most of my props to my dad. He was my mentor, my hero," I said while tracing my path on to college and my early years at NASA before securing what I called "the ultimate gig" as an astronaut. I talked about overcoming the accident that almost cost me my dream and finally getting to see our planet from space. I made sure to encourage viewers to be prepared for serendipity and to stay the course. "Sometimes when you don't believe in yourself, other people do," I said, "and they give you a second chance."

Flipboard bills itself as the world's first social magazine and combines information from across the web into one central, accessible site. The editors invited me to record a segment for a video series called My Magazine. On the set, I sat on a stool while a crew of techs and photographers shot me from various angles using an array of cameras, lighting grids, backdrops, and monitors. The end result was a high-tech production worthy of Hollywood in which images from my career, views of Earth from the space station, and footage of the shuttle liftoff appeared to whirl within an outline of my body. "We as a civilization are explorers," my voice-over stated. "We are all educators in some way. We have some knowledge or some skill to give back. It's what fuels my desire to inspire that next generation of explorers."

The Avis campaign was called "What Drives You." Scenes of a car winding along a ribbon through a majestic mountain range dissolved into scenes of me behind the wheel. "I'm on the road probably a couple weeks a month," my voice intoned. "I love driving. Inside that car is like my own little command module; it's a place of escape. What drives me is to help that one child who doesn't believe in his- or herself." I go on to say, "We have setbacks and we have comebacks. Your comebacks are even sweeter when you stay on the path."

In each campaign I emphasized the importance of inspiring the next generation of explorers. I continued to address that theme in 2015, during which I gave more than twenty talks. While I spoke at places like the Boys and Girls Club and National 4H, my efforts weren't solely focused on young audiences. I spoke to corporations, school systems, and places like the National Science Teachers Conference and the U.S. Air Force. Just as I wanted to motivate students, I also wanted to share my story of grit and determination to empower teachers and organizations to rouse their own constituencies, whether they were young people, employees, or members of professional groups and trade organizations.

Conferences and conventions can be loud and bustling scenes. Even so, they can seem as hushed as a library when compared to the *BattleBots* set. I joined the cast of the ABC show in May 2015, signing on as one of three competition judges. Intense and high-volume, each *BattleBots* episode begins with pulsing futuristic sounds and a bellowing ring announcer, Faruq. The arena is astir with whirling lights and a roaring crowd that gets even louder when Faruq declares, "It's robot fighting time!" Welcome to a world "where rocket scientists are rock stars and girls can be more destructive than boys." Remote-controlled, homemade robots decked out with fighting flippers, powerful hammers, and other weaponry fight to the finish in an arena called the battle box. The winner must win five fights in all, bringing its creators a cash prize and the coveted *BattleBots* trophy, known as the Giant Nut. When a three-minute battle ends with both bots functioning, the judges choose the winner. Split decisions posed the stiffest challenges for us because we had to decide which bot demonstrated the most aggression, inflicted the most damage, pursued the most effective strategy, and exercised the most control.

BattleBots is the ultimate incarnation of experiential real-world, hands-on STEAM learning. In addition to design skills, the roboticists must employ strategy to modify their machines to be more elusive, aggressive, or maneuverable, depending on what their opponents' strengths and weaknesses are. The design aspect of each bot allows for an element of showmanship and artistic flair. My fellow judges Fon Davis (a concept designer and model maker for many Hollywood productions) and Jessica Chobot (host of *Nerdist News* on Nerdist.com) were talented and great to work with. Both *Child Genius* and *BattleBots* illustrated the kinds of skills that were critical to my success as an astronaut. The former required book smarts while the latter demanded grit, guts, and street savvy. The unpredictable nature of spaceflight sometimes requires the crew to shift rapidly from one skill set to the other. If the toilet breaks in space, you can't call the plumber. You have to fix it yourself.

• • •

Alan Ladwig, the emcee at my retirement party, had gone to work for Zero Gravity Corporation before returning to NASA. Zero G uses a specially modified Boeing 727 to create a weightless environment. For a fee, private citizens can book a reservation on the flights and experience zero gravity as the plane executes a series of parabolic arcs. Alan worked at Zero G with an astrophysicist named Maraia Hoffman. He introduced us soon after my retirement party.

Maraia told me about her dream of establishing an extensive facility that would enable citizens to go beyond just experiencing weightlessness to preparing for actual space travel. Her vision intrigued me: "To create a more positive world by enabling people to gain access to space and thereby the Orbital Perspective, which we believe has the power to make a significant impact on our civilization." I had been given this opportunity to fly and it profoundly changed me. I joined her new venture, Star Harbor Space Training Academy, in June 2014, a few months after retiring from NASA. My good friend and astronaut Ron Garan wrote the book *The Orbital Perspective: Lessons in Seeing the Big Picture from a Journey of 71 Million Miles* to help codify what we feel as astronauts in space, and he is part of the team. Star Harbor is going to help many get their Orbital Perspective while on the ground.

In the meantime, I continue to share my tale of grit, grace, and second chances. As I travel around the globe, giving speeches and making friends, Jeannette Williamson Suarez's prophecy is seldom far from my thoughts.

Many things have happened since that day in 2001 when she looked me in the eye and told me about ordeals I'd suffer and victories I'd achieve. I have overcome setbacks and received timely assists from people who saw something in me even when I hadn't seen it myself. Each act of encouragement, each word of kindness, pointed me toward experiences that only a few can claim. My friends and mentors might not have articulated their vision of my potential as forcefully as Jeannette had done, but in the end their language and deeds had been equally prophetic. Each had helped me rise above my limitations, understand my "why," defy gravity, and behold our planet in all its rich and complicated splendor. They gave me the fortitude to return and tell my story, my testimony to the world.

• • •

Tonight, Gracie and I sat on the nursing home porch looking up at the pale translucent satellite—a half moon.

"Mom, we should go there because you won't need your wheelchair."

"OK," she said.

"Three days there, three days back, and one to bounce on the surface. A week's vacation and we'll be home."

Mom turned to me, laughed, and said, "Let's Go."

ACKNOWLEDGMENTS

t has been said that "the two most important days in your life are the day you were born and the day you find out why."

At thirty-seven years old I was very blessed and thankful for Jeanette Suarez to have the courage to share with me my "why"—to share my testimony with the world. The conversation actually started with "something is going to happen to you." Those were heavy, almost unbelievable words to hear when I was in the prime of my life, getting ready to reach for the stars. But that five-minute conversation was the driving force behind my writing *Chasing Space*. There are so many people who've helped guide me on this circuitous fifty-three-year journey, and they kept my path full of positivity, grace, and purpose.

I want to thank my agent, Darnell Strom, at Creative Artists Agency, for routinely encouraging me to tell this story. Thanks also to David Larabell for his help in connecting me with the wonderful Tracy Sherrod at Amistad and David Linker at HarperCollins Children's Books, who both believed in this project from the start. Many thanks to Simon Sinek for many things, but especially for introducing me to Laurie Flynn, who helped get the process started with the proposal. Jabari Asim and Doug Lyons: You helped shape and craft these words, and I appreciate your passion for rich, powerful storytelling.

My family has tirelessly supported me in all my endeavors, no matter how foolish or far out they seemed. Thanks, Mom, Dad, Cat, Allen, Britt, and Ciara. You have always been there for me, and I thank you immensely for your love, encouragement, and support in so many ways. My other family—Louise, Betty, Tom, Chandler, Phyllis, Nanette, Stephanie, Rhonda Ann, Nina, Michael, Branch, Colethia, Cora, Jack, Helen, Anita, Alan, Karen, Harold, Jordon, Renita, Kiera, Geneva, Reggie, Vincent, Arnold, Arnold Jr., Brett, Gregory, Freddie, Rosie, Reen, James, Gladys, Henry, Chanté, and others—I thank you. Thanks for neighbors like the Davises, the Smiths, the Kaines, the Mabrys, the Lynches, the Joneses, the Alexanders, the Brews, the Browns, the Watsons, the Fleshmans, the Bollings, the Powells, and the Saunderses for getting me back on the straight and narrow when I strayed. (Butch, remember you were straying with me.)

Perrymont Elementary educators: Thanks to Principal Carwile, for including me in the sixth-grade class you were teaching algebra to during your break time. That early instruction prepared me for careers in the sciences and in engineering. Educators Sutherland, Bergman, Martin, White, Fowler, Rivers, and others were truly instrumental in my foundational development, and I appreciate your dedication and patience. Stan, Kevbo, Duke, Mike, Butch, Stan W., Rophenia, Brandon, J. W. Charade: You were true friends who helped me stay grounded. Educators at Dunbar Middle—Dot Swain, Leah Ingram, Charlie Dawson, Coach Austin, Cobb, Coach Z, and others—kept the poet in me until this day.

To the Heritage High educators and coaches—I really appreciate you for believing in me. Green, Knight, Storm, Pultz, Gilbert, Farmer, Thomas, Glover, Jones, Hawley, Coleman, Ratliffe, Patterson, Campbell, Pittas, Davis, Switzer, Clarke, Spencer, Mr. Mark, and others, I thank you immensely.

My Jackson Street church family—especially the Williams, the Hutchersons, the Clarkes, the Moselys, and other families—I thank you for helping me keep the faith through many times of turmoil and strife. Jerome, Thad, Alvin, and Calvin—I really appreciate you letting me try to hang with the cool college guys.

My University of Richmond football family—I really appreciate all you did to help me develop a strong sense of grit as we overcame so many "L"s on the gridiron.

Coach Shealy, Coach Hout, Coach Shannon, Coach Van Arsdale, T-lack, Bobby B. Hasty, Worrel, "Red Shoes" Gray, roomie Dan Fitz, Napole, Billy Starke, Don Miller, Damon, Gary O, Cal Bell, J.B., Joe "Betty," and the Jarvis receivers—Jeff, Pup, J.C., John Henry, Doug Ehlers, Johnnie E., and K. J.—were all instrumental in helping me understand what it meant to be on a team, especially during the tough times that first year.

I had a really tough start at the University of Richmond, but there were many people who would not allow me to give up. Dr. Myers (Ad Astra)—you and your family have always been there, along with Drs. Clough, Goldman, Bell, and Dominey. I appreciate the lessons that did not pertain just to chemistry but that helped me to learn about life. Dr. Heilman, John Roush, and the other leaders who did not give up on Spider football made it possible for my NFL stint and prepared me for setbacks and success in space. You also allowed me to see there was more to UR than what I had experienced early on. I can really appreciate what the university has become and how it creates leaders who make significant contributions to our world.

Thanks, Allyn and Lyle, for being great sounding boards and friends during the Lions' camp.

Thanks to my UVA family for helping me get a graduate degree while trying to play in the NFL and believing that I could do both. Glenn Stoner, Ray Taylor, George Cahen, Marlene, and B.J.—you guys kept the notes coming so that I would not get too far behind. To the Killer Bees, thanks for keeping the summers festive—Peggy M., Norwood, Brian, Rodney, Anita, William W., and A.B.C. at the "diss" house.

My NASA family in Hampton really helped me to understand what it meant to be a scientist and an engineer and part of a team that wanted to change the world in a positive way through STEAM. It was also a family of people who believed in uplifting the marginalized and disenfranchised to help them aspire to do great things.

Thanks to Rudy King, John Simmons, Kendall Freeman, "Big Bill," Sam James, Ted Johnson, Joe Heyman, Thomas Kashangaki, David Shannon, Todd Pilot, Charlie Camarda, Rosa Webster, Woodrow Whitlow, Bob Lee, Katherine Johnson, Glinda Shipman, Joe Heyman, Bill Winfree, Bob Rogowski, Elliot Kramer, Meng Cho-Wu, Brooks Childers, Jason, Mark Froggatt, and many others.

My NASA Johnson family took teamwork to another level, especially when we came together to honor our fallen. All my Penguin classmates, my STS-122 and STS-129 crewmates, I thank you for helping make me a better astro and person. Godspeed to our buddies who have left the planet: STS-107

(Rick, Willie, David, Kalpana, Michael, Laurel, and Ilan), Dex, and Piers. Garrett Reisman, thanks so much for being a great CACO who kept me laughing even though I could not hear anything you were saying. Shep and G., thanks for an amazing experience in Star City with Kenny, Bob C., Peggy, Julie, Sandy, and the rest of the Russian Crusader bunch. Dennis, Alla, Sergei, Yuri, thank you for inviting me into your homes and sharing your culture, language, and great food with me. The rest of the CB family as well as the training team: You got us ready and supported us in every way to make sure that we were all safe. Erlinda, Beth, and Heather, you controlled my life for ten years and then were there for my family too. Flight Med John Locke and Jon Clark, you fought for me to get back on flight status, and I appreciate your efforts. Dr. Rich Williams, I thank you for helping the team get comfortable with me flying in space. Joe Dervay, Smith, Brenda, Amy, C.J., Carole: Some of you saw me as an astronaut hopeful and were there to help me get through the labyrinth of forms, documents, and testing to get back on flight status and then get to space; I thank you.

I worked at NASA HQ, which involved two stints that connected me with extremely passionate educators who made a difference in the lives of our most precious resource, our children. Thanks, Charles Scales, for helping me see past my résumé and what could be possible as a different kind of AA for education. The Summer of Innovation crew—Dovie, Carol, Rick, Shelly, Jim, and others—we really helped our kids enjoy STEM over the summer in a hands-on, experiential way. I appreciate all the conferences, symposiums, and gatherings that exposed so many to what we do to share our incredible scientific missions to learners, educators, and the community. As

AA, I was really fortunate to work with this incredible team across the agency. Thanks, Education Coordinating Council. Thanks, CoSTEM—especially Joan F. and others for trying to take a coordinated approach to STEM ed. I learned a lot from many of you. Thanks, Carolyn, Roosevelt, Mea, Donald, Carl, Mabel, Mary, Diane, Joeletta, Lenell, and Andrew for keeping me sane.

Friends, family, colleagues, and complete strangers have helped me through this meandering journey thus far: Paris L., Coach Baber, Coach Russell, Jeanette E., Elizabeth, Pharrell, Christian M., Quincy, Beth N.C., Contessa, Curtis, Robert F., Chopper, Jeanette, Doc Alford, Nancy, Jim and Jere S., Yvonne, the Odyssey team, Barrington, Laura R., Dana, Nicole, Annette, Tsipi, Zigi, Gela, M. Kagan, Bill Nye, Neil DT, Venus, Karen S., Ashley B., Patty O', Sonya D., Olivia, will.i.am, Claudia, Summer, Paul, Ruth, Yolanda, Sherri, Doc Gross, Shaun, Sue, Tom, Willie W., Ed Dwight, Pat S., Tawana, Leah, and Reggie M. Thanks, Melissa, for taking care of the boys so many times, being a good friend, and helping with the edits. The Star Harbor crew—Maraia, Scott, Ron, Mindy, Jenn, Shubham, Luis, Holly, Robert, Alan, and Bill—thanks for having a vision to make things better on this planet by creating an overview for all to see the possibilities. Thank you, Team Constellation: Guy, Ron Anousheh, Nicole, Cora, Jeremy, Jacob, and Christoph.

Finally, Charlie Bolden, I thank you for being a great boss and also like a second father to me. You inspired me and so many others to believe and achieve great things.

I really always tried to do the things I enjoyed but did not always connect a purpose to it. All of you have helped me use my "why" to tell this story, and I appreciate your support and love throughout the years.

ABOUT THE AUTHOR

A former wide receiver for the Detroit Lions, LELAND MELVIN is an engineer and NASA astronaut. He served on the space shuttle *Atlantis* as a mission specialist and was named the NASA Associate Administrator for Education in October 2010. He also served as the cochair on the White House's Federal Coordination in Science, Technology, Engineering, and Mathematics (STEM) Education Task Force, developing the nation's five-year STEM education plan. He is the host of the Lifetime show *Child Genius* and a judge for ABC's *BattleBots*. He holds four honorary doctorates and has received the NFL Player Association Award of Excellence. He lives in Lynchburg, Virginia.